George A. Anastassiou

Intelligent Systems: Approximation by Artificial Neural Networks

T0180969

Intelligent Systems Reference Library, Volume 19

Editors-in-Chief

Prof. Janusz Kacprzyk
Systems Research Institute
Polish Academy of Sciences
ul. Newelska 6
01-447 Warsaw
Poland
E-mail: kacprzyk@ibspan.waw.pl

Prof. Lakhmi C. Jain
University of South Australia
Adelaide
Mawson Lakes Campus
South Australia 5095
Australia
E-mail: Lakhmi.jain@unisa.edu.au

Further volumes of this series can be found on our
homepage: springer.com

Vol. 1. Christine L. Mumford and Lakhmi C. Jain (Eds.)
*Computational Intelligence: Collaboration, Fusion
and Emergence*, 2009
ISBN 978-3-642-01798-8

Vol. 2. Yuehui Chen and Ajith Abraham
*Tree-Structure Based Hybrid
Computational Intelligence*, 2009
ISBN 978-3-642-04738-1

Vol. 3. Anthony Finn and Steve Scheding
*Developments and Challenges for
Autonomous Unmanned Vehicles*, 2010
ISBN 978-3-642-10703-0

Vol. 4. Lakhmi C. Jain and Chee Peng Lim (Eds.)
*Handbook on Decision Making: Techniques
and Applications*, 2010
ISBN 978-3-642-13638-2

Vol. 5. George A. Anastassiou
Intelligent Mathematics: Computational Analysis, 2010
ISBN 978-3-642-17097-3

Vol. 6. Ludmila Dymowa
Soft Computing in Economics and Finance, 2011
ISBN 978-3-642-17718-7

Vol. 7. Gerasimos G. Rigatos
Modelling and Control for Intelligent Industrial Systems,
2011
ISBN 978-3-642-17874-0

Vol. 8. Edward H.Y. Lim, James N.K. Liu, and
Raymond S.T. Lee
*Knowledge Seeker – Ontology Modelling for Information
Search and Management*, 2011
ISBN 978-3-642-17915-0

Vol. 9. Menahem Friedman and Abraham Kandel
Calculus Light, 2011
ISBN 978-3-642-17847-4

Vol. 10. Andreas Tolk and Lakhmi C. Jain
Intelligence-Based Systems Engineering, 2011
ISBN 978-3-642-17930-3

Vol. 11. Samuli Niiranen and Andre Ribeiro (Eds.)
Information Processing and Biological Systems, 2011
ISBN 978-3-642-19620-1

Vol. 12. Florin Gorunescu
Data Mining, 2011
ISBN 978-3-642-19720-8

Vol. 13. Witold Pedrycz and Shyi-Ming Chen (Eds.)
Granular Computing and Intelligent Systems, 2011
ISBN 978-3-642-19819-9

Vol. 14. George A. Anastassiou and Oktay Duman
*Towards Intelligent Modeling: Statistical Approximation
Theory*, 2011
ISBN 978-3-642-19825-0

Vol. 15. Antonino Freno and Edmondo Trentin
Hybrid Random Fields, 2011
ISBN 978-3-642-20307-7

Vol. 16. Alexiei Dingli
*Knowledge Annotation: Making Implicit Knowledge
Explicit*, 2011
ISBN 978-3-642-20322-0

Vol. 17. Crina Grosan and Ajith Abraham
Intelligent Systems, 2011
ISBN 978-3-642-21003-7

Vol. 18. Achim Zielesny
From Curve Fitting to Machine Learning, 2011
ISBN 978-3-642-21279-6

Vol. 19. George A. Anastassiou
*Intelligent Systems: Approximation by Artificial Neural
Networks*, 2011
ISBN 978-3-642-21430-1

George A. Anastassiou

Intelligent Systems: Approximation by Artificial Neural Networks

 Springer

Prof. George A. Anastassiou
University of Memphis
Department of Mathematical Sciences
Memphis, TN 38152
USA
E-mail: ganastss@memphis.edu

ISBN 978-3-642-26855-7   ISBN 978-3-642-21431-8 (eBook)

DOI 10.1007/978-3-642-21431-8

Intelligent Systems Reference Library   ISSN 1868-4394

Typeset & Cover Design: Scientific Publishing Services Pvt. Ltd., Chennai, India.

Printed on acid-free paper

9 8 7 6 5 4 3 2 1

springer.com

Preface

This brief monograph is the first one to deal exclusively with the quantitative approximation by artificial neural networks to the identity-unit operator. Here we study with rates the approximation properties of the "right" sigmoidal and hyperbolic tangent artificial neural network positive linear operators. In particular we study the degree of approximation of these operators to the unit operator in the univariate and multivariate cases over bounded or unbounded domains. This is given via inequalities and with the use of modulus of continuity of the involved function or its higher order derivative. We examine the real and complex cases.

For the convenience of the reader, the chapters of this book are written in a self-contained style.

This treatise relies on author's last two years of related research work.

Advanced courses and seminars can be taught out of this brief book. All necessary background and motivations are given per chapter. A related list of references is given also per chapter. My book's results appeared for the first time in my published articles which are mentioned throughout the references. They are expected to find applications in many areas of computer science and applied mathematics, such as neural networks, intelligent systems, complexity theory, learning theory, vision and approximation theory, etc. As such this monograph is suitable for researchers, graduate students, and seminars of the above subjects, also for all science libraries.

The preparation of this booklet took place during 2010-2011 in Memphis, Tennessee, USA.

I would like to thank my family for their dedication and love to me, which was the strongest support during the writing of this book.

March 1, 2011   George A. Anastassiou
                                        Department of Mathematical Sciences
                                          The University of Memphis, TN, U.S.A.

Contents

Contents

Chapter 1
Univariate Sigmoidal Neural Network Quantitative Approximation

Here we give the univariate quantitative approximation of real and complex valued continuous functions on a compact interval or all the real line by quasi-interpolation sigmoidal neural network operators. This approximation is obtained by establishing Jackson type inequalities involving the modulus of continuity of the engaged function or its high order derivative. The operators are defined by using a density function induced by the logarithmic sigmoidal function. Our approximations are pointwise and with respect to the uniform norm. The related feed-forward neural network is with one hidden layer. This chapter relies on [4].

1.1 Introduction

Feed-forward neural networks (FNNs) with one hidden layer, the only type of networks we deal with in this chapter, are mathematically expressed as

$$N_n\left(x\right) = \sum_{j=0}^{n} c_j \sigma\left(\langle a_j \cdot x\rangle + b_j\right), \quad x \in \mathbb{R}^s, \quad s \in \mathbb{N},$$

where for $0 \leq j \leq n$, $b_j \in \mathbb{R}$ are the thresholds, $a_j \in \mathbb{R}^s$ are the connection weights, $c_j \in \mathbb{R}$ are the coefficients, $\langle a_j \cdot x\rangle$ is the inner product of a_j and x, and σ is the activation function of the network. In many fundamental network models, the activation function is the sigmoidal function of logistic type.

It is well known that FNNs are universal approximators. Theoretically, any continuous function defined on a compact set can be approximated to any desired degree of accuracy by increasing the number of hidden neurons. It was shown by Cybenko [11] and Funahashi [13], that any continuous function

G.A. Anastassiou: Intelligent Systems: Approximation by ANN, ISRL 19, pp. 1–32.
springerlink.com

can be approximated on a compact set with uniform topology by a network of the form $N_n(x)$, using any continuous, sigmoidal activation function. Hornik et al. in [15], have proved that any measurable function can be approached with such a network. Furthermore, these authors established in [16], that any function of the Sobolev spaces can be approached with all derivatives. A variety of density results on FNN approximations to multivariate functions were later established by many authors using different methods, for more or less general situations: [18] by Leshno et al., [22] by Mhaskar and Micchelli, [10] by Chui and Li, [8] by Chen and Chen, [14] by Hahm and Hong, etc.

Usually these results only give theorems about the existence of an approximation. A related and important problem is that of complexity: determining the number of neurons required to guarantee that all functions belonging to a space can be approximated to the prescribed degree of accuracy ϵ.

Barron [5] shows that if the function is assumed to satisfy certain conditions expressed in terms of its Fourier transform, and if each of the neurons evaluates a sigmoidal activation function, then at most $O\left(\epsilon^{-2}\right)$ neurons are needed to achieve the order of approximation ϵ. Some other authors have published similar results on the complexity of FNN approximations: Mhaskar and Micchelli [23], Suzuki [24], Maiorov and Meir [20], Makovoz [21], Ferrari and Stengel [12], Xu and Cao [26], Cao et al. [7], etc.

The author in [1] and [2], see chapters 2-5, was the first to establish neural network approximations to continuous functions with rates by very specifically defined neural network operators of Cardaliagnet-Euvrard and "Squashing" types, by employing the modulus of continuity of the engaged function or its high order derivative, and producing very tight Jackson type inequalities. He treats there both the univariate and multivariate cases. The defining these operators "bell-shaped" and "squashing" function are assumed to be of compact support. Also in [2] he gives the Nth order asymptotic expansion for the error of weak approximation of these two operators to a special natural class of smooth functions, see chapters 4-5 there.

For this chapter the author is greatly motivated by the important article [9] by Z. Chen and F. Cao.

He presents related to it work and much more beyond however [9] remains the initial point. So the author here performs univariate sigmoidal neural network approximations with rates to continuous functions over compact intervals of the real line or over the whole $\mathbb{R}$, then he extends his results to complex valued functions. All convergences here are with rates expressed via the modulus of continuity of the involved function or its high order derivative, and given by very tight Jackson type inequalities.

The author presents here the correct and precisely defined quasi-interpolation neural network operator related to compact intervals, and among others, improves results from [9]. The compact intervals are not necessarily symmetric to the origin. Some of the upper bounds to error quantity are very flexible and general. In preparation to establish our results we prove further properties of the basic density function defining our operators.

1.2 Background and Auxiliary Results

We consider here the sigmoidal function of logarithmic type

$$s(x) = \frac{1}{1 + e^{-x}}, \quad x \in \mathbb{R}.$$

It has the properties $\lim_{x \to +\infty} s(x) = 1$ and $\lim_{x \to -\infty} s(x) = 0$.

This function plays the role of an activation function in the hidden layer of neural networks, also has application in biology, demography, etc. ([6, 17]).

As in [9], we consider

$$\Phi(x) := \frac{1}{2} \left(s(x+1) - s(x-1) \right), \quad x \in \mathbb{R}.$$

It has the following properties:

i) $\Phi(x) > 0$, $\forall\, x \in \mathbb{R}$,

ii) $\sum_{k=-\infty}^{\infty} \Phi(x - k) = 1$, $\forall\, x \in \mathbb{R}$,

iii) $\sum_{k=-\infty}^{\infty} \Phi(nx - k) = 1$, $\forall\, x \in \mathbb{R}; n \in \mathbb{N}$,

iv) $\int_{-\infty}^{\infty} \Phi(x)\, dx = 1$,

v) Φ is a density function,

vi) Φ is even: $\Phi(-x) = \Phi(x)$, $x \geq 0$.

We observe that ([9])

$$\Phi(x) = \left(\frac{e^2 - 1}{2e} \right) \frac{e^{-x}}{(1 + e^{-x-1})(1 + e^{-x+1})} = \left(\frac{e^2 - 1}{2e^2} \right) \frac{1}{(1 + e^{x-1})(1 + e^{-x-1})},$$

and

$$\Phi'(x) = \left(\frac{e^2 - 1}{2e^2} \right) \left[-\frac{(e^x - e^{-x})}{e(1 + e^{x-1})^2 (1 + e^{-x-1})^2} \right] \leq 0, \ x \geq 0.$$

Hence

vii) Φ is decreasing on $\mathbb{R}_+$, and increasing on $\mathbb{R}_-$.

Let $0 < \alpha < 1$, $n \in \mathbb{N}$. We see the following

$$\sum_{\left\{ \substack{k = -\infty \\ : |nx - k| > n^{1-\alpha}} \right\}}^{\infty} \Phi(nx - k) = \sum_{\left\{ \substack{k = -\infty \\ : |nx - k| > n^{1-\alpha}} \right\}}^{\infty} \Phi(|nx - k|) \le$$

$$\left(\frac{e^2 - 1}{2e^2} \right) \int_{(n^{1-\alpha}-1)}^{\infty} \frac{1}{(1 + e^{x-1})(1 + e^{-x-1})} dx \le$$

$$\left(\frac{e^2 - 1}{2e} \right) \int_{(n^{1-\alpha}-1)}^{\infty} e^{-x} dx = \left(\frac{e^2 - 1}{2e} \right) \left(e^{-(n^{1-\alpha}-1)} \right)$$

$$= \left(\frac{e^2 - 1}{2} \right) e^{-n^{(1-\alpha)}} = 3.1992 e^{-n^{(1-\alpha)}}.$$

We have found that:

viii) for $n \in \mathbb{N}$, $0 < \alpha < 1$, we get

$$\sum_{\left\{ \substack{k = -\infty \\ : |nx - k| > n^{1-\alpha}} \right\}}^{\infty} \Phi(nx - k) < \left(\frac{e^2 - 1}{2} \right) e^{-n^{(1-\alpha)}} = 3.1992 e^{-n^{(1-\alpha)}}.$$

Denote by $\lceil \cdot \rceil$ the ceiling of a number, and by $\lfloor \cdot \rfloor$ the integral part of a number. Consider $x \in [a, b] \subset \mathbb{R}$ and $n \in \mathbb{N}$ such that $\lceil na \rceil \le \lfloor nb \rfloor$.

We observe that

$$1 = \sum_{k=-\infty}^{\infty} \Phi(nx - k) > \sum_{k=\lceil na \rceil}^{\lfloor nb \rfloor} \Phi(nx - k) =$$

$$\sum_{k=\lceil na \rceil}^{\lfloor nb \rfloor} \Phi(|nx - k|) > \Phi(|nx - k_0|),$$

for any $k_0 \in [\lceil na \rceil, \lfloor nb \rfloor] \cap \mathbb{Z}$.

Here we can choose $k_0 \in [\lceil na \rceil, \lfloor nb \rfloor] \cap \mathbb{Z}$ such that $|nx - k_0| < 1$.

Therefore $\Phi\left(|nx - k_0|\right) > \Phi\left(1\right) = 0.19046485$.
Consequently,

$$\sum_{k=\lceil na \rceil}^{\lfloor nb \rfloor} \Phi\left(nx - k\right) > \Phi\left(1\right) = 0.19046485.$$

Therefore we get

ix) $\dfrac{1}{\sum_{k=\lceil na \rceil}^{\lfloor nb \rfloor} \Phi(nx-k)} < \dfrac{1}{\Phi(1)} = 5.250312578, \ \forall \ x \in [a, b]$.

We also notice that

$$1 - \sum_{k=\lceil na \rceil}^{\lfloor nb \rfloor} \Phi\left(nb - k\right) = \sum_{k=-\infty}^{\lceil na \rceil - 1} \Phi\left(nb - k\right) + \sum_{k=\lfloor nb \rfloor + 1}^{\infty} \Phi\left(nb - k\right)$$

$$> \Phi\left(nb - \lfloor nb \rfloor - 1\right)$$

(call $\varepsilon := nb - \lfloor nb \rfloor, \ 0 \le \varepsilon < 1$)

$$= \Phi\left(\varepsilon - 1\right) = \Phi\left(1 - \varepsilon\right) \ge \Phi\left(1\right) > 0.$$

Therefore $\lim_{n \to \infty} \left(1 - \sum_{k=\lceil na \rceil}^{\lfloor nb \rfloor} \Phi\left(nb - k\right)\right) > 0$.
Similarly,

$$1 - \sum_{k=\lceil na \rceil}^{\lfloor nb \rfloor} \Phi\left(na - k\right) = \sum_{k=-\infty}^{\lceil na \rceil - 1} \Phi\left(na - k\right) + \sum_{k=\lfloor nb \rfloor + 1}^{\infty} \Phi\left(na - k\right)$$

$$> \Phi\left(na - \lceil na \rceil + 1\right)$$

(call $\eta := \lceil na \rceil - na, \ 0 \le \eta < 1$)

$$= \Phi\left(1 - \eta\right) \ge \Phi\left(1\right) > 0.$$

Therefore again $\lim_{n \to \infty} \left(1 - \sum_{k=\lceil na \rceil}^{\lfloor nb \rfloor} \Phi\left(na - k\right)\right) > 0$.
Therefore we obtain that

x) $\lim_{n \to \infty} \sum_{k=\lceil na \rceil}^{\lfloor nb \rfloor} \Phi\left(nx - k\right) \ne 1$, for at least some $x \in [a, b]$.

Let $f \in C\left([a, b]\right)$ and $n \in \mathbb{N}$ such that $\lceil na \rceil \le \lfloor nb \rfloor$.
We introduce and define the positive linear neural network operator

$$G_n\left(f, x\right) := \dfrac{\sum_{k=\lceil na \rceil}^{\lfloor nb \rfloor} f\left(\frac{k}{n}\right) \Phi\left(nx - k\right)}{\sum_{k=\lceil na \rceil}^{\lfloor nb \rfloor} \Phi\left(nx - k\right)}, \quad x \in [a.b]. \tag{1.1}$$

For large enough n we always have $\lceil na \rceil \leq \lfloor nb \rfloor$. Also $a \leq \frac{k}{n} \leq b$, iff $\lceil na \rceil \leq k \leq \lfloor nb \rfloor$.

We study here the pointwise convergence of $G_n(f, x)$ to $f(x)$ with rates.
For convenience we call

$$G_n^*(f, x) := \sum_{k=\lceil na \rceil}^{\lfloor nb \rfloor} f\left(\frac{k}{n}\right) \Phi(nx - k), \tag{1.2}$$

that is

$$G_n(f, x) := \frac{G_n^*(f, x)}{\sum_{k=\lceil na \rceil}^{\lfloor nb \rfloor} \Phi(nx - k)}. \tag{1.3}$$

Thus,

$$G_n(f, x) - f(x) = \frac{G_n^*(f, x)}{\sum_{k=\lceil na \rceil}^{\lfloor nb \rfloor} \Phi(nx - k)} - f(x)$$

$$= \frac{G_n^*(f, x) - f(x) \sum_{k=\lceil na \rceil}^{\lfloor nb \rfloor} \Phi(nx - k)}{\sum_{k=\lceil na \rceil}^{\lfloor nb \rfloor} \Phi(nx - k)}. \tag{1.4}$$

Consequently we derive

$$|G_n(f, x) - f(x)| \leq \frac{1}{\Phi(1)} \left| G_n^*(f, x) - f(x) \sum_{k=\lceil na \rceil}^{\lfloor nb \rfloor} \Phi(nx - k) \right|. \tag{1.5}$$

That is

$$|G_n(f, x) - f(x)| \leq (5.250312578) \left| \sum_{k=\lceil na \rceil}^{\lfloor nb \rfloor} \left(f\left(\frac{k}{n}\right) - f(x) \right) \Phi(nx - k) \right|. \tag{1.6}$$

We will estimate the right hand side of (1.6).

For that we need, for $f \in C([a, b])$ the first modulus of continuity

$$\omega_1(f, h) := \sup_{\substack{x, y \in [a, b] \\ |x - y| \leq h}} |f(x) - f(y)|, \quad h > 0. \tag{1.7}$$

Similarly it is defined for $f \in C_B(\mathbb{R})$ (continuous and bounded on $\mathbb{R}$). We have that $\lim_{h \to 0} \omega_1(f, h) = 0$.

When $f \in C_B(\mathbb{R})$ we define, (see also [9])

$$\overline{G}_n(f, x) := \sum_{k=-\infty}^{\infty} f\left(\frac{k}{n}\right) \Phi(nx - k), \quad n \in \mathbb{N}, \, x \in \mathbb{R}, \tag{1.8}$$

the quasi-interpolation neural network operator.

By [3] we derive the following three theorems on extended Taylor formula.

Theorem 1.1. *Let* $N \in \mathbb{N}$, $0 < \varepsilon < \frac{\pi}{2}$ *small, and* $f \in C^N \left(\left[-\frac{\pi}{2} + \varepsilon, \frac{\pi}{2} - \varepsilon \right] \right)$; $x, y \in \left[-\frac{\pi}{2} + \varepsilon, \frac{\pi}{2} - \varepsilon \right]$. *Then*

$$f(x) = f(y) + \sum_{k=1}^{N} \frac{\left(f \circ \sin^{-1} \right)^{(k)} (\sin y)}{k!} (\sin x - \sin y)^k + K_N(y, x), \quad (1.9)$$

where

$$K_N(y, x) = \frac{1}{(N-1)!}. \quad (1.10)$$

$$\int_y^x (\sin x - \sin s)^{N-1} \left(\left(f \circ \sin^{-1} \right)^{(N)} (\sin s) - \left(f \circ \sin^{-1} \right)^{(N)} (\sin y) \right) \cos s \, ds.$$

Theorem 1.2. *Let* $f \in C^N \left([\varepsilon, \pi - \varepsilon] \right)$, $N \in \mathbb{N}$, $\varepsilon > 0$ *small;* $x, y \in [\varepsilon, \pi - \varepsilon]$. *Then*

$$f(x) = f(y) + \sum_{k=1}^{N} \frac{\left(f \circ \cos^{-1} \right)^{(k)} (\cos y)}{k!} (\cos x - \cos y)^k + K_N^*(y, x), \quad (1.11)$$

where

$$K_N^*(y, x) = -\frac{1}{(N-1)!}. \quad (1.12)$$

$$\int_y^x (\cos x - \cos s)^{N-1} \left[\left(f \circ \cos^{-1} \right)^{(N)} (\cos s) - \left(f \circ \cos^{-1} \right)^{(N)} (\cos y) \right] \sin s \, ds.$$

Theorem 1.3. *Let* $f \in C^N \left([a, b] \right)$ *(or* $f \in C^N (\mathbb{R})$*),* $N \in \mathbb{N}$; $x, y \in [a, b]$ *(or* $x, y \in \mathbb{R}$*). Then*

$$f(x) = f(y) + \sum_{k=1}^{N} \frac{\left(f \circ \ln_{\frac{1}{e}} \right)^{(k)} (e^{-y})}{k!} \left(e^{-x} - e^{-y} \right)^k + \overline{K}_N(y, x), \quad (1.13)$$

where

$$\overline{K}_N(y, x) = -\frac{1}{(N-1)!}. \quad (1.14)$$

$$\int_y^x \left(e^{-x} - e^{-s} \right)^{N-1} \left[\left(f \circ \ln_{\frac{1}{e}} \right)^{(N)} \left(e^{-s} \right) - \left(f \circ \ln_{\frac{1}{e}} \right)^{(N)} \left(e^{-y} \right) \right] e^{-s} ds.$$

Remark 1.4. *Using the mean value theorem we get*

$$|\sin x - \sin y| \leq |x - y|, \quad (1.15)$$
$$|\cos x - \cos y| \leq |x - y|, \quad \forall \, x, y \in \mathbb{R},$$

furthermore we have

$$|\sin x - \sin y| \leq 2, \quad (1.16)$$
$$|\cos x - \cos y| \leq 2, \quad \forall \, x, y \in \mathbb{R}.$$

Similarly we have

$$\left|e^{-x} - e^{-y}\right| \le e^{-a} \left|x - y\right|, \tag{1.17}$$

and

$$\left|e^{-x} - e^{-y}\right| \le e^{-a} - e^{-b}, \quad \forall \, x, y \in [a, b]. \tag{1.18}$$

Let $g(x) = \ln_{\frac{1}{e}} x$, $\sin^{-1} x$, $\cos^{-1} x$ and assume $f^{(j)}(x_0) = 0$, $k = 1, ..., N$. Then, by [3], we get $\left(f \circ g^{-1}\right)^{(j)}(g(x_0)) = 0$, $j = 1, ..., N$.

Remark 1.5. *It is well known that* $e^x > x^m$, $m \in \mathbb{N}$, *for large* $x > 0$.
Let fixed $\alpha, \beta > 0$, *then* $\left\lceil \frac{\alpha}{\beta} \right\rceil \in \mathbb{N}$, *and for large* $x > 0$ *we have*

$$e^x > x^{\left\lceil \frac{\alpha}{\beta} \right\rceil} \ge x^{\frac{\alpha}{\beta}}.$$

So for suitable very large $x > 0$ *we obtain*

$$e^{x^\beta} > \left(x^\beta\right)^{\frac{\alpha}{\beta}} = x^\alpha.$$

We proved for large $x > 0$ *and* $\alpha, \beta > 0$ *that*

$$e^{x^\beta} > x^\alpha. \tag{1.19}$$

Therefore for large $n \in \mathbb{N}$ *and fixed* $\alpha, \beta > 0$, *we have*

$$e^{n^\beta} > n^\alpha. \tag{1.20}$$

That is

$$e^{-n^\beta} < n^{-\alpha}, \quad \text{for large } n \in \mathbb{N}. \tag{1.21}$$

So for $0 < \alpha < 1$ *we get*

$$e^{-n^{(1-\alpha)}} < n^{-\alpha}. \tag{1.22}$$

Thus be given fixed $A, B > 0$, for the linear combination $\left(An^{-\alpha} + Be^{-n^{(1-\alpha)}}\right)$ the (dominant) rate of convergence to zero is $n^{-\alpha}$.

The closer α is to 1 we get faster and better rate of convergence to zero.

1.3   Real Neural Network Quantitative Approximations

Here we present a series of neural network approximations to a function given with rates.

We first give.

Theorem 1.6. *Let $f \in C([a,b])$, $0 < \alpha < 1$, $n \in \mathbb{N}$, $x \in [a,b]$. Then*
i)

$$|G_n(f,x) - f(x)| \le (5.250312578)\left[\omega_1\left(f, \frac{1}{n^\alpha}\right) + 6.3984\|f\|_\infty\, e^{-n^{(1-\alpha)}}\right] =: \lambda,$$
(1.23)

and
ii)

$$\|G_n(f) - f\|_\infty \le \lambda,$$
(1.24)

where $\|\cdot\|_\infty$ is the supremum norm.

Proof. We observe that

$$\left|\sum_{k=\lceil na\rceil}^{\lfloor nb\rfloor}\left(f\left(\frac{k}{n}\right) - f(x)\right)\Phi(nx-k)\right| \le$$

$$\sum_{k=\lceil na\rceil}^{\lfloor nb\rfloor}\left|f\left(\frac{k}{n}\right) - f(x)\right|\Phi(nx-k) =$$

$$\sum_{\substack{k=\lceil na\rceil \\ \left|\frac{k}{n}-x\right|\le \frac{1}{n^\alpha}}}^{\lfloor nb\rfloor}\left|f\left(\frac{k}{n}\right) - f(x)\right|\Phi(nx-k) +$$

$$\sum_{\substack{k=\lceil na\rceil \\ \left|\frac{k}{n}-x\right|> \frac{1}{n^\alpha}}}^{\lfloor nb\rfloor}\left|f\left(\frac{k}{n}\right) - f(x)\right|\Phi(nx-k) \le$$

$$\sum_{\substack{k=\lceil na\rceil \\ \left|\frac{k}{n}-x\right|\le \frac{1}{n^\alpha}}}^{\lfloor nb\rfloor}\omega_1\left(f,\left|\frac{k}{n}-x\right|\right)\Phi(nx-k) +$$

$$2\|f\|_\infty\sum_{\substack{k=\lceil na\rceil \\ |k-nx|> n^{1-\alpha}}}^{\lfloor nb\rfloor}\Phi(nx-k) \le$$

$$\omega_1\left(f,\frac{1}{n^\alpha}\right)\sum_{\substack{k=-\infty \\ \left|\frac{k}{n}-x\right|\le \frac{1}{n^\alpha}}}^{\infty}\Phi(nx-k) +$$

$$2 \left\| f \right\|_\infty \sum_{\substack{k = -\infty \\ |k - nx| > n^{1-\alpha}}}^{\infty} \Phi(nx - k) \underset{\text{(by (viii))}}{\leq}$$

$$\omega_1 \left(f, \frac{1}{n^\alpha} \right) + 2 \left\| f \right\|_\infty (3.1992) \, e^{-n^{(1-\alpha)}}$$

$$= \omega_1 \left(f, \frac{1}{n^\alpha} \right) + 6.3984 \left\| f \right\|_\infty e^{-n^{(1-\alpha)}}.$$

That is

$$\left| \sum_{k=\lceil na \rceil}^{\lfloor nb \rfloor} \left(f \left(\frac{k}{n} \right) - f(x) \right) \Phi(nx - k) \right| \leq \omega_1 \left(f, \frac{1}{n^\alpha} \right) + 6.3984 \left\| f \right\|_\infty e^{-n^{(1-\alpha)}}.$$

Using (1.6) we prove the claim. ∎

Theorem 1.6 improves a lot the model of neural network approximation of [9], see Theorem 3 there.

Next we give

Theorem 1.7. *Let* $f \in C_B(\mathbb{R})$, $0 < \alpha < 1$, $n \in \mathbb{N}$, $x \in \mathbb{R}$. *Then*
i)

$$\left| \overline{G}_n(f, x) - f(x) \right| \leq \omega_1 \left(f, \frac{1}{n^\alpha} \right) + 6.3984 \left\| f \right\|_\infty e^{-n^{(1-\alpha)}} =: \mu, \qquad (1.25)$$

and
ii)

$$\left\| \overline{G}_n(f) - f \right\|_\infty \leq \mu. \qquad (1.26)$$

Proof. We see that

$$\left| \overline{G}_n(f, x) - f(x) \right| = \left| \sum_{k=-\infty}^{\infty} f \left(\frac{k}{n} \right) \Phi(nx - k) - f(x) \sum_{k=-\infty}^{\infty} \Phi(nx - k) \right| =$$

$$\left| \sum_{k=-\infty}^{\infty} \left(f \left(\frac{k}{n} \right) - f(x) \right) \Phi(nx - k) \right| \leq \sum_{k=-\infty}^{\infty} \left| f \left(\frac{k}{n} \right) - f(x) \right| \Phi(nx - k) =$$

$$\sum_{\substack{k = -\infty \\ \left| \frac{k}{n} - x \right| \leq \frac{1}{n^\alpha}}}^{\infty} \left| f \left(\frac{k}{n} \right) - f(x) \right| \Phi(nx - k) +$$

$$\sum_{\substack{k = -\infty \\ \left| \frac{k}{n} - x \right| > \frac{1}{n^\alpha}}}^{\infty} \left| f \left(\frac{k}{n} \right) - f(x) \right| \Phi(nx - k) \leq$$

$$\sum_{\substack{k=-\infty \\ \left|\frac{k}{n} - x\right| \le \frac{1}{n^\alpha}}}^{\infty} \omega_1\left(f, \left|\frac{k}{n} - x\right|\right) \Phi(nx - k) +$$

$$2\|f\|_\infty \sum_{\substack{k=-\infty \\ \left|\frac{k}{n} - x\right| > \frac{1}{n^\alpha}}}^{\infty} \Phi(nx - k) \le$$

$$\omega_1\left(f, \frac{1}{n^\alpha}\right) \sum_{\substack{k=-\infty \\ \left|\frac{k}{n} - x\right| \le \frac{1}{n^\alpha}}}^{\infty} \Phi(nx - k) +$$

$$2\|f\|_\infty \sum_{\substack{k=-\infty \\ |k - nx| > n^{1-\alpha}}}^{\infty} \Phi(nx - k) \underset{\text{(by (viii))}}{\le}$$

$$\omega_1\left(f, \frac{1}{n^\alpha}\right) + 2\|f\|_\infty (3.1992) e^{-n^{(1-\alpha)}}$$

$$= \omega_1\left(f, \frac{1}{n^\alpha}\right) + 6.3984 \|f\|_\infty e^{-n^{(1-\alpha)}},$$

proving the claim. ∎

Theorem 1.7 improves Theorem 4 of [9].

In the next we discuss high order of approximation by using the smoothness of f.

Theorem 1.8. *Let* $f \in C^N([a, b])$, $n, N \in \mathbb{N}$, $0 < \alpha < 1$, $x \in [a, b]$. *Then*
i)

$$|G_n(f, x) - f(x)| \le (5.250312578) \cdot \tag{1.27}$$

$$\left\{ \sum_{j=1}^{N} \frac{|f^{(j)}(x)|}{j!} \left[\frac{1}{n^{\alpha j}} + (3.1992)(b - a)^j e^{-n^{(1-\alpha)}} \right] + \right.$$

$$\left. \left[\omega_1\left(f^{(N)}, \frac{1}{n^\alpha}\right) \frac{1}{n^{\alpha N} N!} + \frac{(6.3984)\|f^{(N)}\|_\infty (b - a)^N}{N!} e^{-n^{(1-\alpha)}} \right] \right\},$$

ii) *assume further* $f^{(j)}(x_0) = 0$, $j = 1, ..., N$, *for some* $x_0 \in [a, b]$, *it holds*

$$|G_n(f, x_0) - f(x_0)| \le (5.250312578) \cdot \tag{1.28}$$

$$\left[\omega_1\left(f^{(N)}, \frac{1}{n^\alpha}\right) \frac{1}{n^{\alpha N} N!} + \frac{(6.3984)\|f^{(N)}\|_\infty (b - a)^N}{N!} e^{-n^{(1-\alpha)}} \right],$$

notice here the extremely high rate of convergence at $n^{-(N+1)\alpha}$,
iii)

$$\|G_n(f) - f\|_\infty \le (5.250312578) \cdot \tag{1.29}$$

$$\left\{ \sum_{j=1}^{N} \frac{\|f^{(j)}\|_\infty}{j!} \left[\frac{1}{n^{\alpha j}} + (3.1992)(b-a)^j e^{-n^{(1-\alpha)}} \right] + \right.$$

$$\left. \left[\omega_1\left(f^{(N)}, \frac{1}{n^\alpha}\right) \frac{1}{n^{\alpha N} N!} + \frac{(6.3984)\|f^{(N)}\|_\infty (b-a)^N}{N!} e^{-n^{(1-\alpha)}} \right] \right\}.$$

Proof. Next we apply Taylor's formula with integral remainder.
We have (here $\frac{k}{n}, x \in [a,b]$)

$$f\left(\frac{k}{n}\right) = \sum_{j=0}^{N} \frac{f^{(j)}(x)}{j!} \left(\frac{k}{n} - x\right)^j + \int_x^{\frac{k}{n}} \left(f^{(N)}(t) - f^{(N)}(x)\right) \frac{\left(\frac{k}{n} - t\right)^{N-1}}{(N-1)!} dt.$$

Then

$$f\left(\frac{k}{n}\right) \Phi(nx - k) = \sum_{j=0}^{N} \frac{f^{(j)}(x)}{j!} \Phi(nx - k) \left(\frac{k}{n} - x\right)^j +$$

$$\Phi(nx - k) \int_x^{\frac{k}{n}} \left(f^{(N)}(t) - f^{(N)}(x)\right) \frac{\left(\frac{k}{n} - t\right)^{N-1}}{(N-1)!} dt.$$

Hence

$$\sum_{k=\lceil na \rceil}^{\lfloor nb \rfloor} f\left(\frac{k}{n}\right) \Phi(nx - k) - f(x) \sum_{k=\lceil na \rceil}^{\lfloor nb \rfloor} \Phi(nx - k) =$$

$$\sum_{j=1}^{N} \frac{f^{(j)}(x)}{j!} \sum_{k=\lceil na \rceil}^{\lfloor nb \rfloor} \Phi(nx - k) \left(\frac{k}{n} - x\right)^j +$$

$$\sum_{k=\lceil na \rceil}^{\lfloor nb \rfloor} \Phi(nx - k) \int_x^{\frac{k}{n}} \left(f^{(N)}(t) - f^{(N)}(x)\right) \frac{\left(\frac{k}{n} - t\right)^{N-1}}{(N-1)!} dt.$$

Thus

$$G_n^*(f, x) - f(x) \left(\sum_{k=\lceil na \rceil}^{\lfloor nb \rfloor} \Phi(nx - k) \right) = \sum_{j=1}^{N} \frac{f^{(j)}(x)}{j!} G_n^* \left((\cdot - x)^j \right) + \Lambda_n(x),$$

where

$$\Lambda_n(x) := \sum_{k=\lceil na \rceil}^{\lfloor nb \rfloor} \Phi(nx - k) \int_x^{\frac{k}{n}} \left(f^{(N)}(t) - f^{(N)}(x)\right) \frac{\left(\frac{k}{n} - t\right)^{N-1}}{(N-1)!} dt.$$

We suppose that $b - a > \frac{1}{n^\alpha}$, which is always the case for large enough $n \in \mathbb{N}$, that is when $n > \left\lceil (b-a)^{-\frac{1}{\alpha}} \right\rceil$.

Thus $\left| \frac{k}{n} - x \right| \leq \frac{1}{n^\alpha}$ or $\left| \frac{k}{n} - x \right| > \frac{1}{n^\alpha}$.
As in [2], pp. 72-73 for

$$\gamma := \int_x^{\frac{k}{n}} \left(f^{(N)}(t) - f^{(N)}(x) \right) \frac{\left(\frac{k}{n} - t \right)^{N-1}}{(N-1)!} dt,$$

in case of $\left| \frac{k}{n} - x \right| \leq \frac{1}{n^\alpha}$, we find that

$$|\gamma| \leq \omega_1 \left(f^{(N)}, \frac{1}{n^\alpha} \right) \frac{1}{n^{\alpha N} N!}$$

(for $x \leq \frac{k}{n}$ or $x \geq \frac{k}{n}$).

Notice also for $x \leq \frac{k}{n}$ that

$$\left| \int_x^{\frac{k}{n}} \left(f^{(N)}(t) - f^{(N)}(x) \right) \frac{\left(\frac{k}{n} - t \right)^{N-1}}{(N-1)!} dt \right| \leq$$

$$\int_x^{\frac{k}{n}} \left| f^{(N)}(t) - f^{(N)}(x) \right| \frac{\left(\frac{k}{n} - t \right)^{N-1}}{(N-1)!} dt \leq$$

$$2 \left\| f^{(N)} \right\|_\infty \int_x^{\frac{k}{n}} \frac{\left(\frac{k}{n} - t \right)^{N-1}}{(N-1)!} dt = 2 \left\| f^{(N)} \right\|_\infty \frac{\left(\frac{k}{n} - x \right)^N}{N!} \leq 2 \left\| f^{(N)} \right\|_\infty \frac{(b-a)^N}{N!}.$$

Next assume $\frac{k}{n} \leq x$, then

$$\left| \int_x^{\frac{k}{n}} \left(f^{(N)}(t) - f^{(N)}(x) \right) \frac{\left(\frac{k}{n} - t \right)^{N-1}}{(N-1)!} dt \right| =$$

$$\left| \int_{\frac{k}{n}}^x \left(f^{(N)}(t) - f^{(N)}(x) \right) \frac{\left(\frac{k}{n} - t \right)^{N-1}}{(N-1)!} dt \right| \leq$$

$$\int_{\frac{k}{n}}^x \left| f^{(N)}(t) - f^{(N)}(x) \right| \frac{\left(t - \frac{k}{n} \right)^{N-1}}{(N-1)!} dt \leq$$

$$2 \left\| f^{(N)} \right\|_\infty \int_{\frac{k}{n}}^x \frac{\left(t - \frac{k}{n} \right)^{N-1}}{(N-1)!} dt = 2 \left\| f^{(N)} \right\|_\infty \frac{\left(x - \frac{k}{n} \right)^N}{N!} \leq 2 \left\| f^{(N)} \right\|_\infty \frac{(b-a)^N}{N!}.$$

Thus

$$|\gamma| \leq 2 \left\| f^{(N)} \right\|_\infty \frac{(b-a)^N}{N!},$$

in all two cases.

Therefore

$$\Lambda_n(x) = \sum_{\substack{k=\lceil na\rceil \\ \left|\frac{k}{n}-x\right|\le \frac{1}{n^\alpha}}}^{\lfloor nb\rfloor} \Phi(nx-k)\gamma + \sum_{\substack{k=\lceil na\rceil \\ \left|\frac{k}{n}-x\right|> \frac{1}{n^\alpha}}}^{\lfloor nb\rfloor} \Phi(nx-k)\gamma.$$

Hence

$$|\Lambda_n(x)| \le \sum_{\substack{k=\lceil na\rceil \\ \left|\frac{k}{n}-x\right|\le \frac{1}{n^\alpha}}}^{\lfloor nb\rfloor} \Phi(nx-k)\left(\omega_1\left(f^{(N)},\frac{1}{n^\alpha}\right)\frac{1}{N!n^{N\alpha}}\right) +$$

$$\left(\sum_{\substack{k=\lceil na\rceil \\ \left|\frac{k}{n}-x\right|> \frac{1}{n^\alpha}}}^{\lfloor nb\rfloor} \Phi(nx-k)\right) 2\left\|f^{(N)}\right\|_\infty \frac{(b-a)^N}{N!} \le$$

$$\omega_1\left(f^{(N)},\frac{1}{n^\alpha}\right)\frac{1}{N!n^{N\alpha}} + 2\left\|f^{(N)}\right\|_\infty \frac{(b-a)^N}{N!}(3.1992)\,e^{-n^{(1-\alpha)}}.$$

Consequently we have

$$|\Lambda_n(x)| \le \omega_1\left(f^{(N)},\frac{1}{n^\alpha}\right)\frac{1}{n^{\alpha N}N!} + (6.3984)\frac{\left\|f^{(N)}\right\|_\infty (b-a)^N}{N!}e^{-n^{(1-\alpha)}}.$$

We further see that

$$G_n^*\left((\cdot-x)^j\right) = \sum_{k=\lceil na\rceil}^{\lfloor nb\rfloor} \Phi(nx-k)\left(\frac{k}{n}-x\right)^j.$$

Therefore

$$\left|G_n^*\left((\cdot-x)^j\right)\right| \le \sum_{k=\lceil na\rceil}^{\lfloor nb\rfloor} \Phi(nx-k)\left|\frac{k}{n}-x\right|^j =$$

$$\sum_{\substack{k=\lceil na\rceil \\ \left|\frac{k}{n}-x\right|\le \frac{1}{n^\alpha}}}^{\lfloor nb\rfloor} \Phi(nx-k)\left|\frac{k}{n}-x\right|^j + \sum_{\substack{k=\lceil na\rceil \\ \left|\frac{k}{n}-x\right|> \frac{1}{n^\alpha}}}^{\lfloor nb\rfloor} \Phi(nx-k)\left|\frac{k}{n}-x\right|^j \le$$

$$\frac{1}{n^{\alpha j}}\sum_{\substack{k=\lceil na\rceil \\ \left|\frac{k}{n}-x\right|\le \frac{1}{n^\alpha}}}^{\lfloor nb\rfloor} \Phi(nx-k) + (b-a)^j \sum_{\substack{k=\lceil na\rceil \\ |k-nx|> n^{1-\alpha}}}^{\lfloor nb\rfloor} \Phi(nx-k)$$

$$\le \frac{1}{n^{\alpha j}} + (b-a)^j \, (3.1992) \, e^{-n^{(1-\alpha)}}.$$

Hence

$$\left| G_n^* \left((\cdot - x)^j \right) \right| \le \frac{1}{n^{\alpha j}} + (b-a)^j \, (3.1992) \, e^{-n^{(1-\alpha)}},$$

for $j = 1, ..., N$.

Putting things together we have established

$$|G_n^* (f, x) - f(x)| \le \sum_{j=1}^{N} \frac{\left| f^{(j)}(x) \right|}{j!} \left[\frac{1}{n^{\alpha j}} + (3.1992)(b-a)^j \, e^{-n^{(1-\alpha)}} \right] +$$

$$\left[\omega_1 \left(f^{(N)}, \frac{1}{n^\alpha} \right) \frac{1}{n^{\alpha N} N!} + \frac{(6.3984) \left\| f^{(N)} \right\|_\infty (b-a)^N}{N!} e^{-n^{(1-\alpha)}} \right],$$

that is establishing theorem. ∎

We make

Remark 1.9. *We notice that*

$$G_n (f, x) - \sum_{j=1}^{N} \frac{f^{(j)}(x)}{j!} G_n \left((\cdot - x)^j \right) - f(x) =$$

$$\frac{G_n^* (f, x)}{\left(\sum_{k=\lceil na \rceil}^{\lfloor nb \rfloor} \Phi(nx - k) \right)} - \frac{1}{\left(\sum_{k=\lceil na \rceil}^{\lfloor nb \rfloor} \Phi(nx - k) \right)} \left(\sum_{j=1}^{n} \frac{f^{(j)}(x)}{j!} G_n^* \left((\cdot - x)^j \right) \right)$$

$$- f(x) = \frac{1}{\left(\sum_{k=\lceil na \rceil}^{\lfloor nb \rfloor} \Phi(nx - k) \right)} \cdot$$

$$\left[G_n^* (f, x) - \left(\sum_{j=1}^{n} \frac{f^{(j)}(x)}{j!} G_n^* \left((\cdot - x)^j \right) \right) - \left(\sum_{k=\lceil na \rceil}^{\lfloor nb \rfloor} \Phi(nx - k) \right) f(x) \right].$$

Therefore we get

$$\left| G_n (f, x) - \sum_{j=1}^{N} \frac{f^{(j)}(x)}{j!} G_n \left((\cdot - x)^j \right) - f(x) \right| \le (5.250312578) \cdot$$

$$\left| G_n^* (f, x) - \left(\sum_{j=1}^{n} \frac{f^{(j)}(x)}{j!} G_n^* \left((\cdot - x)^j \right) \right) - \left(\sum_{k=\lceil na \rceil}^{\lfloor nb \rfloor} \Phi(nx - k) \right) f(x) \right|,$$

$$\tag{1.30}$$

$\forall \, x \in [a, b].$

In the next three Theorems 1.10-1.12 we present more general and flexible upper bounds to our error quantities.

We give

Theorem 1.10. *Let* $f \in C^N\left([a,b]\right)$, $n, N \in \mathbb{N}$, $0 < \alpha < 1$, $x \in [a,b]$. *Then*
1)

$$\left| G_n\left(f,x\right) - \sum_{j=1}^{N} \frac{\left(f \circ \ln_{\frac{1}{e}}\right)^{(j)}\left(e^{-x}\right)}{j!} G_n\left(\left(e^{-\cdot} - e^{-x}\right)^j, x\right) - f\left(x\right) \right| \leq$$

$$(5.250312578)\left[\frac{e^{-aN}}{N! n^{N\alpha}} \omega_1\left(\left(f \circ \ln_{\frac{1}{e}}\right)^{(N)}, \frac{e^{-a}}{n^\alpha}\right) + \right.$$

$$\left. \frac{(6.3984)\left(e^{-a} - e^{-b}\right)^N}{N!} \left\| \left(f \circ \ln_{\frac{1}{e}}\right)^{(N)} \right\|_\infty e^{-n^{(1-\alpha)}} \right], \qquad (1.31)$$

2)

$$|G_n\left(f,x\right) - f\left(x\right)| \leq (5.250312578) \cdot \qquad (1.32)$$

$$\left\{ \sum_{j=1}^{N} \frac{\left| \left(f \circ \ln_{\frac{1}{e}}\right)^{(j)}\left(e^{-x}\right) \right|}{j!} e^{-aj}\left[\frac{1}{n^{\alpha j}} + (3.1992)(b-a)^j e^{-n^{(1-\alpha)}} \right] + \right.$$

$$\left[\frac{e^{-aN}}{N! n^{\alpha N}} \omega_1\left(\left(f \circ \ln_{\frac{1}{e}}\right)^{(N)}, \frac{e^{-a}}{n^\alpha}\right) + \right.$$

$$\left. \left. \frac{(6.3984)\left(e^{-a} - e^{-b}\right)^N}{N!} \left\| \left(f \circ \ln_{\frac{1}{e}}\right)^{(N)} \right\|_\infty e^{-n^{(1-\alpha)}} \right] \right\},$$

3) *If* $f^{(j)}\left(x_0\right) = 0$, $j = 1, ..., N$, *it holds*

$$|G_n\left(f,x_0\right) - f\left(x_0\right)| \leq (5.250312578) \cdot \qquad (1.33)$$

$$\left[\frac{e^{-aN}}{N! n^{N\alpha}} \omega_1\left(\left(f \circ \ln_{\frac{1}{e}}\right)^{(N)}, \frac{e^{-a}}{n^\alpha}\right) + \right.$$

$$\left. \frac{(6.3984)\left(e^{-a} - e^{-b}\right)^N}{N!} \left\| \left(f \circ \ln_{\frac{1}{e}}\right)^{(N)} \right\|_\infty e^{-n^{(1-\alpha)}} \right].$$

Observe here the speed of convergence is extremely high at $\frac{1}{n^{(N+1)\alpha}}$.

Proof. Call $F := f \circ \ln_{\frac{1}{e}}$. Let $x, \frac{k}{n} \in [a,b]$. Then

$$f\left(\frac{k}{n}\right) - f\left(x\right) = \sum_{j=1}^{N} \frac{F^{(j)}\left(e^{-x}\right)}{j!}\left(e^{-\frac{k}{n}} - e^{-x}\right)^j + \overline{K}_N\left(x, \frac{k}{n}\right),$$

where

$$\overline{K}_N\left(x, \frac{k}{n}\right) := -\frac{1}{(N-1)!} \cdot$$

$$\int_x^{\frac{k}{n}} \left(e^{-\frac{k}{n}} - e^{-s}\right)^{N-1} \left[F^{(N)}\left(e^{-s}\right) - F^{(N)}\left(e^{-x}\right)\right] e^{-s} ds.$$

Thus

$$\sum_{k=\lceil na \rceil}^{\lfloor nb \rfloor} \Phi\left(nx - k\right) f\left(\frac{k}{n}\right) - f\left(x\right) \sum_{k=\lceil na \rceil}^{\lfloor nb \rfloor} \Phi\left(nx - k\right) =$$

$$\sum_{j=1}^{N} \frac{F^{(j)}\left(e^{-x}\right)}{j!} \sum_{k=\lceil na \rceil}^{\lfloor nb \rfloor} \Phi\left(nx - k\right) \left(e^{-\frac{k}{n}} - e^{-x}\right)^j + \sum_{k=\lceil na \rceil}^{\lfloor nb \rfloor} \Phi\left(nx - k\right) \frac{1}{(N-1)!} \cdot$$

$$\int_x^{\frac{k}{n}} \left(e^{-\frac{k}{n}} - e^{-s}\right)^{N-1} \left[F^{(N)}\left(e^{-s}\right) - F^{(N)}\left(e^{-x}\right)\right] de^{-s}.$$

Therefore

$$G_n^*\left(f, x\right) - f\left(x\right) \left(\sum_{k=\lceil na \rceil}^{\lfloor nb \rfloor} \Phi\left(nx - k\right)\right) =$$

$$\sum_{j=1}^{N} \frac{F^{(j)}\left(e^{-x}\right)}{j!} G_n^*\left(\left(e^{-\cdot} - e^{-x}\right)^j, x\right) + U_n\left(x\right),$$

where

$$U_n\left(x\right) := \sum_{k=\lceil na \rceil}^{\lfloor nb \rfloor} \Phi\left(nx - k\right) \mu,$$

with

$$\mu := \frac{1}{(N-1)!} \int_{e^{-x}}^{e^{-\frac{k}{n}}} \left(e^{-\frac{k}{n}} - w\right)^{N-1} \left[F^{(N)}\left(w\right) - F^{(N)}\left(e^{-x}\right)\right] dw.$$

Case of $\left|\frac{k}{n} - x\right| \leq \frac{1}{n^\alpha}$.
i) Subcase of $x \geq \frac{k}{n}$. I.e. $e^{-\frac{k}{n}} \geq e^{-x}$.

$$|\mu| \leq \frac{1}{(N-1)!} \int_{e^{-x}}^{e^{-\frac{k}{n}}} \left(e^{-\frac{k}{n}} - w\right)^{N-1} \left|F^{(N)}\left(w\right) - F^{(N)}\left(e^{-x}\right)\right| dw \leq$$

$$\frac{1}{(N-1)!} \int_{e^{-x}}^{e^{-\frac{k}{n}}} \left(e^{-\frac{k}{n}} - w\right)^{N-1} \omega_1\left(F^{(N)}, \left|w - e^{-x}\right|\right) dw \leq$$

$$\frac{1}{(N-1)!} \omega_1\left(F^{(N)}, \left|e^{-\frac{k}{n}} - e^{-x}\right|\right) \int_{e^{-x}}^{e^{-\frac{k}{n}}} \left(e^{-\frac{k}{n}} - w\right)^{N-1} dw \leq$$

$$\frac{\left(e^{-\frac{k}{n}} - e^{-x}\right)^N}{N!} \omega_1\left(F^{(N)}, e^{-a} \left|x - \frac{k}{n}\right|\right) \leq$$

$$e^{-aN} \frac{\left| x - \frac{k}{n} \right|^N}{N!} \omega_1 \left(F^{(N)}, e^{-a} \left| x - \frac{k}{n} \right| \right) \leq$$

$$\frac{e^{-aN}}{N! n^{\alpha N}} \omega_1 \left(F^{(N)}, \frac{e^{-a}}{n^\alpha} \right).$$

Hence when $x \geq \frac{k}{n}$ we have

$$|\mu| \leq \frac{e^{-aN}}{N! n^{\alpha N}} \omega_1 \left(F^{(N)}, \frac{e^{-a}}{n^\alpha} \right).$$

ii) Subcase of $\frac{k}{n} \geq x$. Then $e^{-\frac{k}{n}} \leq e^{-x}$ and

$$|\mu| \leq \frac{1}{(N-1)!} \int_{e^{-\frac{k}{n}}}^{e^{-x}} \left(w - e^{-\frac{k}{n}} \right)^{N-1} \left| F^{(N)}(w) - F^{(N)}\left(e^{-x} \right) \right| dw \leq$$

$$\frac{1}{(N-1)!} \int_{e^{-\frac{k}{n}}}^{e^{-x}} \left(w - e^{-\frac{k}{n}} \right)^{N-1} \omega_1 \left(F^{(N)}, \left| w - e^{-x} \right| \right) dw \leq$$

$$\frac{1}{(N-1)!} \omega_1 \left(F^{(N)}, e^{-x} - e^{-\frac{k}{n}} \right) \int_{e^{-\frac{k}{n}}}^{e^{-x}} \left(w - e^{-\frac{k}{n}} \right)^{N-1} dw \leq$$

$$\frac{1}{(N-1)!} \omega_1 \left(F^{(N)}, e^{-a} \left| x - \frac{k}{n} \right| \right) \frac{\left(e^{-x} - e^{-\frac{k}{n}} \right)^N}{N} \leq$$

$$\frac{1}{N!} \omega_1 \left(F^{(N)}, \frac{e^{-a}}{n^\alpha} \right) e^{-aN} \left| x - \frac{k}{n} \right|^N \leq$$

$$\frac{1}{N!} \omega_1 \left(F^{(N)}, \frac{e^{-a}}{n^\alpha} \right) \frac{e^{-aN}}{n^{\alpha N}},$$

i.e.

$$|\mu| \leq \frac{e^{-aN}}{N! n^{\alpha N}} \omega_1 \left(F^{(N)}, \frac{e^{-a}}{n^\alpha} \right),$$

when $\frac{k}{n} \geq x$. So in general when $\left| \frac{k}{n} - x \right| \leq \frac{1}{n^\alpha}$ we proved that

$$|\mu| \leq \frac{e^{-aN}}{N! n^{\alpha N}} \omega_1 \left(F^{(N)}, \frac{e^{-a}}{n^\alpha} \right).$$

Also we observe:

i)' When $\frac{k}{n} \leq x$, we get

$$|\mu| \leq \frac{1}{(N-1)!} \left(\int_{e^{-x}}^{e^{-\frac{k}{n}}} \left(e^{-\frac{k}{n}} - w \right)^{N-1} dw \right) 2 \left\| F^{(N)} \right\|_\infty$$

$$= \frac{2\left\|F^{(N)}\right\|_\infty}{(N-1)!} \frac{\left(e^{-\frac{k}{n}} - e^{-x}\right)^N}{N} \le \frac{2\left\|F^{(N)}\right\|_\infty}{N!} \left(e^{-a} - e^{-b}\right)^N.$$

ii)' When $\frac{k}{n} \ge x$, we obtain

$$|\mu| \le \frac{1}{(N-1)!} \left(\int_{e^{-\frac{k}{n}}}^{e^{-x}} \left(w - e^{-\frac{k}{n}}\right)^{N-1} dw\right) 2\left\|F^{(N)}\right\|_\infty$$

$$= \frac{2\left\|F^{(N)}\right\|_\infty}{N!} \left(e^{-x} - e^{-\frac{k}{n}}\right)^N \le \frac{2\left\|F^{(N)}\right\|_\infty}{N!} \left(e^{-a} - e^{-b}\right)^N.$$

We proved always true that

$$|\mu| \le \frac{2\left(e^{-a} - e^{-b}\right)^N \left\|F^{(N)}\right\|_\infty}{N!}.$$

Consequently we find

$$|U_n(x)| \le \sum_{\substack{k = \lceil na \rceil \\ \left|\frac{k}{n} - x\right| \le \frac{1}{n^\alpha}}}^{\lfloor nb \rfloor} \Phi(nx - k)|\mu| + \sum_{\substack{k = \lceil na \rceil \\ \left|\frac{k}{n} - x\right| > \frac{1}{n^\alpha}}}^{\lfloor nb \rfloor} \Phi(nx - k)|\mu| \le$$

$$\left(\frac{e^{-aN}}{N!n^{\alpha N}}\omega_1\left(F^{(N)}, \frac{e^{-a}}{n^\alpha}\right)\right) \left(\sum_{k=-\infty}^\infty \Phi(nx - k)\right) +$$

$$\left(\frac{2\left(e^{-a} - e^{-b}\right)^N \left\|F^{(N)}\right\|_\infty}{N!}\right) \left(\sum_{\substack{k = -\infty \\ |k - nx| > n^{1-\alpha}}}^\infty \Phi(nx - k)\right) \le$$

$$\frac{e^{-aN}}{N!n^{\alpha N}}\omega_1\left(F^{(N)}, \frac{e^{-a}}{n^\alpha}\right) + \left(\frac{2\left(e^{-a} - e^{-b}\right)^N \left\|F^{(N)}\right\|_\infty}{N!}\right) (3.1992) e^{-n^{(1-\alpha)}}.$$

So we have proved that

$$|U_n(x)| \le \frac{e^{-aN}}{N!n^{\alpha N}}\omega_1\left(F^{(N)}, \frac{e^{-a}}{n^\alpha}\right) + \frac{(6.3984)\left(e^{-a} - e^{-b}\right)^N \left\|F^{(N)}\right\|_\infty}{N!} e^{-n^{(1-\alpha)}}.$$

We also notice that

$$\left|G_n^*\left(\left(e^{-\cdot} - e^{-x}\right)^j, x\right)\right| \le G_n^*\left(\left|e^{-\cdot} - e^{-x}\right|^j, x\right) \le$$

$$e^{-aj} G_n^* \left(|\cdot - x|^j , x \right) = e^{-aj} \left(\sum_{k=\lceil na \rceil}^{\lfloor nb \rfloor} \Phi (nx - k) \left| \frac{k}{n} - x \right|^j \right) = e^{-aj}.$$

$$\left[\sum_{\substack{k=\lceil na \rceil \\ |x-\frac{k}{n}| \le \frac{1}{n^\alpha}}}^{\lfloor nb \rfloor} \Phi(nx-k) \left| \frac{k}{n} - x \right|^j + \sum_{\substack{k=\lceil na \rceil \\ |x-\frac{k}{n}| > \frac{1}{n^\alpha}}}^{\lfloor nb \rfloor} \Phi(nx-k) \left| \frac{k}{n} - x \right|^j \le \right.$$

$$e^{-aj} \left[\frac{1}{n^{\alpha j}} + (b-a)^j \sum_{\substack{k=\lceil na \rceil \\ |x-\frac{k}{n}| > \frac{1}{n^\alpha}}}^{\lfloor nb \rfloor} \Phi (nx - k) \le \right.$$

$$e^{-aj} \left[\frac{1}{n^{\alpha j}} + (3.1992) (b-a)^j e^{-n^{(1-\alpha)}} \right].$$

Thus we have established

$$\left| G_n^* \left((e^{-\cdot} - e^{-x})^j , x \right) \right| \le e^{-aj} \left[\frac{1}{n^{\alpha j}} + (3.1992) (b-a)^j e^{-n^{(1-\alpha)}} \right],$$

for $j = 1, ..., N$, and the theorem.   ∎

We continue with

Theorem 1.11. *Let* $f \in C^N \left(\left[-\frac{\pi}{2} + \varepsilon, \frac{\pi}{2} - \varepsilon \right] \right)$, $n, N \in \mathbb{N}$, $0 < \varepsilon < \frac{\pi}{2}$, ε *small,* $x \in \left[-\frac{\pi}{2} + \varepsilon, \frac{\pi}{2} - \varepsilon \right]$, $0 < \alpha < 1$. *Then*
1)

$$\left| G_n (f,x) - \sum_{j=1}^{N} \frac{ \left(f \circ \sin^{-1} \right)^{(j)} (\sin x)}{j!} G_n \left((\sin \cdot - \sin x)^j , x \right) - f(x) \right| \le$$

$$\tag{1.34}$$

$$(5.250312578) \left[\frac{\omega_1 \left(\left(f \circ \sin^{-1} \right)^{(N)} , \frac{1}{n^\alpha} \right)}{n^{\alpha N} N!} + \right.$$

$$\left. \left(\frac{(3.1992) \, 2^{N+1} \left\| \left(f \circ \sin^{-1} \right)^{(N)} \right\|_\infty}{N!} \right) e^{-n^{(1-\alpha)}} \right],$$

2)

$$|G_n (f,x) - f(x)| \le (5.250312578) \cdot$$

$$\left\{\sum_{j=1}^{N}\frac{\left|\left(f\circ\sin^{-1}\right)^{(j)}\left(\sin x\right)\right|}{j!}\left[\frac{1}{n^{\alpha j}}+\left(3.1992\right)\left(\pi-2\varepsilon\right)^{j}e^{-n^{(1-\alpha)}}\right]+\right.$$

$$\left.\left[\frac{\omega_1\left(\left(f\circ\sin^{-1}\right)^{(N)},\frac{1}{n^{\alpha}}\right)}{N!n^{\alpha N}}+\left(\frac{\left(3.1992\right)2^{N+1}}{N!}\left\|\left(f\circ\sin^{-1}\right)^{(N)}\right\|_{\infty}\right)e^{-n^{(1-\alpha)}}\right]\right\},$$

$$(1.35)$$

3) assume further $f^{(j)}\left(x_0\right)=0$, $j=1,...,N$ for some $x_0\in\left[-\frac{\pi}{2}+\varepsilon,\frac{\pi}{2}-\varepsilon\right]$, it holds

$$\left|G_n\left(f,x_0\right)-f\left(x_0\right)\right|\le\left(5.250312578\right)\cdot\qquad(1.36)$$

$$\left[\frac{\omega_1\left(\left(f\circ\sin^{-1}\right)^{(N)},\frac{1}{n^{\alpha}}\right)}{n^{\alpha N}N!}+\left(\frac{\left(3.1992\right)2^{N+1}\left\|\left(f\circ\sin^{-1}\right)^{(N)}\right\|_{\infty}}{N!}\right)e^{-n^{(1-\alpha)}}\right].$$

Notice in the last the high speed of convergence of order $n^{-\alpha(N+1)}$.

Proof. Call $F:=f\circ\sin^{-1}$ and let $\frac{k}{n},x\in\left[-\frac{\pi}{2}+\varepsilon,\frac{\pi}{2}-\varepsilon\right]$. Then

$$f\left(\frac{k}{n}\right)-f\left(x\right)=\sum_{j=1}^{N}\frac{F^{(j)}\left(\sin x\right)}{j!}\left(\sin\frac{k}{n}-\sin x\right)^{j}+$$

$$\frac{1}{(N-1)!}\int_{x}^{\frac{k}{n}}\left(\sin\frac{k}{n}-\sin s\right)^{N-1}\left(F^{(N)}\left(\sin s\right)-F^{(N)}\left(\sin x\right)\right)d\sin s.$$

Hence

$$\sum_{k=\lceil na\rceil}^{\lfloor nb\rfloor}f\left(\frac{k}{n}\right)\Phi\left(nx-k\right)-f\left(x\right)\sum_{k=\lceil na\rceil}^{\lfloor nb\rfloor}\Phi\left(nx-k\right)=$$

$$\sum_{j=1}^{N}\frac{F^{(j)}\left(\sin x\right)}{j!}\sum_{k=\lceil na\rceil}^{\lfloor nb\rfloor}\Phi(nx-k)\left(\sin\frac{k}{n}-\sin x\right)^{j}+\frac{1}{(N-1)!}\sum_{k=\lceil na\rceil}^{\lfloor nb\rfloor}\Phi\left(nx-k\right)\cdot$$

$$\int_{x}^{\frac{k}{n}}\left(\sin\frac{k}{n}-\sin s\right)^{N-1}\left(F^{(N)}\left(\sin s\right)-F^{(N)}\left(\sin x\right)\right)d\sin s.$$

Set here $a=-\frac{\pi}{2}+\varepsilon$, $b=\frac{\pi}{2}-\varepsilon$. Thus

$$G_n^{*}\left(f,x\right)-f\left(x\right)\sum_{k=\lceil na\rceil}^{\lfloor nb\rfloor}\Phi\left(nx-k\right)=$$

$$\sum_{j=1}^{N}\frac{F^{(j)}\left(\sin x\right)}{j!}G_n^{*}\left(\left(\sin\cdot-\sin x\right)^{j},x\right)+M_n\left(x\right),$$

where

$$M_n(x) := \sum_{k=\lceil na \rceil}^{\lfloor nb \rfloor} \Phi(nx - k)\, \rho,$$

with

$$\rho := \frac{1}{(N-1)!} \int_x^{\frac{k}{n}} \left(\sin \frac{k}{n} - \sin s\right)^{N-1} \left(F^{(N)}(\sin s) - F^{(N)}(\sin x)\right) d\sin s.$$

Case of $\left|\frac{k}{n} - x\right| \le \frac{1}{n^\alpha}$.

i) Subcase of $\frac{k}{n} \ge x$. The function $\sin$ is increasing on $[a, b]$, i.e. $\sin \frac{k}{n} \ge \sin x$.

Then

$$|\rho| \le \frac{1}{(N-1)!} \int_x^{\frac{k}{n}} \left(\sin \frac{k}{n} - \sin s\right)^{N-1} \left|F^{(N)}(\sin s) - F^{(N)}(\sin x)\right| d\sin s =$$

$$\frac{1}{(N-1)!} \int_{\sin x}^{\sin \frac{k}{n}} \left(\sin \frac{k}{n} - w\right)^{N-1} \left(F^{(N)}(w) - F^{(N)}(\sin x)\right) dw \le$$

$$\frac{1}{(N-1)!} \int_{\sin x}^{\sin \frac{k}{n}} \left(\sin \frac{k}{n} - w\right)^{N-1} \omega_1\left(F^{(N)}, |w - \sin x|\right) dw \le$$

$$\omega_1\left(F^{(N)}, \left|\sin \frac{k}{n} - \sin x\right|\right) \frac{\left(\sin \frac{k}{n} - \sin x\right)^N}{N!} \le$$

$$\omega_1\left(F^{(N)}, \left|\frac{k}{n} - x\right|\right) \frac{\left(\frac{k}{n} - x\right)^N}{N!} \le \omega_1\left(F^{(N)}, \frac{1}{n^\alpha}\right) \frac{1}{n^{\alpha N} N!}.$$

So if $\frac{k}{n} \ge x$, then

$$|\rho| \le \omega_1\left(F^{(N)}, \frac{1}{n^\alpha}\right) \frac{1}{N! n^{\alpha N}}.$$

ii) Subcase of $\frac{k}{n} \le x$, then $\sin \frac{k}{n} \le \sin x$. Hence

$$|\rho| \le \frac{1}{(N-1)!} \int_{\frac{k}{n}}^x \left(\sin s - \sin \frac{k}{n}\right)^{N-1} \left|F^{(N)}(\sin s) - F^{(N)}(\sin x)\right| d\sin s =$$

$$\frac{1}{(N-1)!} \int_{\sin \frac{k}{n}}^{\sin x} \left(w - \sin \frac{k}{n}\right)^{N-1} \left|F^{(N)}(w) - F^{(N)}(\sin x)\right| dw \le$$

$$\frac{1}{(N-1)!} \int_{\sin \frac{k}{n}}^{\sin x} \left(w - \sin \frac{k}{n}\right)^{N-1} \omega_1\left(F^{(N)}, |w - \sin x|\right) dw \le$$

$$\frac{1}{(N-1)!} \omega_1\left(F^{(N)}, \left|\sin x - \sin \frac{k}{n}\right|\right) \int_{\sin \frac{k}{n}}^{\sin x} \left(w - \sin \frac{k}{n}\right)^{N-1} dw \le$$

$$\frac{1}{(N-1)!}\omega_1\left(F^{(N)},\left|x-\frac{k}{n}\right|\right)\frac{\left(\sin x-\sin\frac{k}{k}\right)^N}{N}\leq$$

$$\frac{1}{N!}\omega_1\left(F^{(N)},\frac{1}{n^\alpha}\right)\left|x-\frac{k}{n}\right|^N\leq\frac{1}{n^{\alpha N}N!}\omega_1\left(F^{(N)},\frac{1}{n^\alpha}\right).$$

We got for $\frac{k}{n}\leq x$ that

$$|\rho|\leq\omega_1\left(F^{(N)},\frac{1}{n^\alpha}\right)\frac{1}{n^{\alpha N}N!}.$$

So in both cases we proved that

$$|\rho|\leq\omega_1\left(F^{(N)},\frac{1}{n^\alpha}\right)\frac{1}{n^{\alpha N}N!},$$

when $\left|\frac{k}{n}-x\right|\leq\frac{1}{n^\alpha}$.

Also in general ($\frac{k}{n}\geq x$ case)

$$|\rho|\leq\frac{1}{(N-1)!}\left(\int_x^{\frac{k}{n}}\left(\sin\frac{k}{n}-\sin s\right)^{N-1}d\sin s\right)2\left\|F^{(N)}\right\|_\infty=$$

$$\frac{1}{(N-1)!}\left(\int_{\sin x}^{\sin\frac{k}{n}}\left(\sin\frac{k}{n}-w\right)^{N-1}dw\right)2\left\|F^{(N)}\right\|_\infty=$$

$$\frac{1}{N!}\left(\sin\frac{k}{n}-\sin x\right)^N 2\left\|F^{(N)}\right\|_\infty\leq\frac{2^{N+1}}{N!}\left\|F^{(N)}\right\|_\infty.$$

Also (case of $\frac{k}{n}\leq x$) we get

$$|\rho|\leq\frac{1}{(N-1)!}\left(\int_{\frac{k}{n}}^x\left(\sin s-\sin\frac{k}{n}\right)^{N-1}d\sin s\right)2\left\|F^{(N)}\right\|_\infty=$$

$$\frac{1}{(N-1)!}\left(\int_{\sin\frac{k}{n}}^{\sin x}\left(w-\sin\frac{k}{n}\right)^{N-1}dw\right)2\left\|F^{(N)}\right\|_\infty=$$

$$\frac{1}{(N-1)!}\frac{\left(\sin x-\sin\frac{k}{n}\right)^N}{N}2\left\|F^{(N)}\right\|_\infty\leq\frac{2^{N+1}}{N!}\left\|F^{(N)}\right\|_\infty.$$

So we obtained in general that

$$|\rho|\leq\frac{2^{N+1}}{N!}\left\|F^{(N)}\right\|_\infty.$$

Therefore we derive

$$|M_n(x)|\leq\sum_{\substack{k=\lceil na\rceil\\\left(k:\left|x-\frac{k}{n}\right|\leq\frac{1}{n^\alpha}\right)}}^{\lfloor nb\rfloor}\Phi(nx-k)|\rho|+\sum_{\substack{k=\lceil na\rceil\\\left(k:\left|x-\frac{k}{n}\right|>\frac{1}{n^\alpha}\right)}}^{\lfloor nb\rfloor}\Phi(nx-k)|\rho|\leq$$

$$\left(\frac{\omega_1\left(F^{(N)}, \frac{1}{n^\alpha}\right)}{N! n^{\alpha N}}\right) + \frac{(3.1992)\, 2^{N+1}}{N!} \left\|F^{(N)}\right\|_\infty e^{-n^{(1-\alpha)}}.$$

So that

$$|M_n(x)| \le \frac{\omega_1\left(F^{(N)}, \frac{1}{n^\alpha}\right)}{N! n^{\alpha N}} + \frac{(3.1992)\, 2^{N+1}}{N!} \left\|F^{(N)}\right\|_\infty e^{-n^{(1-\alpha)}}.$$

Next we estimate

$$\left|G_n^*\left((\sin \cdot - \sin x)^j, x\right)\right| \le G_n^*\left(|\sin \cdot - \sin x|^j, x\right) \le$$

$$G_n^*\left(|\cdot - x|^j, x\right) = \sum_{k=\lceil na \rceil}^{\lfloor nb \rfloor} \Phi(nx - k) \left|\frac{k}{n} - x\right|^j \le$$

(work as before)

$$\frac{1}{n^{\alpha j}} + (3.1992)(\pi - 2\varepsilon)^j e^{-n^{(1-\alpha)}}.$$

Therefore

$$\left|G_n^*\left((\sin \cdot - \sin x)^j, x\right)\right| \le \frac{1}{n^{\alpha j}} + (3.1992)(\pi - 2\varepsilon)^j e^{-n^{(1-\alpha)}},$$

$j = 1, ..., N$.

The theorem is proved. ∎

We finally give

Theorem 1.12. *Let* $f \in C^N\left([\varepsilon, \pi - \varepsilon]\right)$, $n, N \in \mathbb{N}$, $\varepsilon > 0$ *small*, $x \in [\varepsilon, \pi - \varepsilon]$, $0 < \alpha < 1$. *Then*

1)

$$\left|G_n(f, x) - \sum_{j=1}^{N} \frac{\left(f \circ \cos^{-1}\right)^{(j)}(\cos x)}{j!} G_n\left((\cos \cdot - \cos x)^j, x\right) - f(x)\right| \le$$

$$(1.37)$$

$$(5.250312578) \left[\frac{\omega_1\left(\left(f \circ \cos^{-1}\right)^{(N)}, \frac{1}{n^\alpha}\right)}{n^{\alpha N} N!} + \right.$$

$$\left. \left(\frac{(3.1992)\, 2^{N+1} \left\|\left(f \circ \cos^{-1}\right)^{(N)}\right\|_\infty}{N!}\right) e^{-n^{(1-\alpha)}}\right],$$

2)

$$|G_n(f, x) - f(x)| \le (5.250312578) \cdot$$

$$\left\{ \sum_{j=1}^{N} \frac{\left|\left(f \circ \cos^{-1}\right)^{(j)} (\cos x)\right|}{j!} \left[\frac{1}{n^{\alpha j}} + (3.1992)(\pi - 2\varepsilon)^j e^{-n^{(1-\alpha)}}\right] + \right.$$

$$\left. \left[\frac{\omega_1 \left(\left(f \circ \cos^{-1}\right)^{(N)}, \frac{1}{n^{\alpha}}\right)}{N! n^{\alpha N}} + \left(\frac{(3.1992) \, 2^{N+1}}{N!} \left\|\left(f \circ \cos^{-1}\right)^{(N)}\right\|_{\infty}\right) e^{-n^{(1-\alpha)}}\right]\right\},$$

$$\tag{1.38}$$

3) *assume further* $f^{(j)}(x_0) = 0$, $j = 1, ..., N$ *for some* $x_0 \in [\varepsilon, \pi - \varepsilon]$, *it holds*

$$|G_n(f, x_0) - f(x_0)| \le (5.250312578) \cdot \tag{1.39}$$

$$\left[\frac{\omega_1 \left(\left(f \circ \cos^{-1}\right)^{(N)}, \frac{1}{n^{\alpha}}\right)}{n^{\alpha N} N!} + \left(\frac{(3.1992) \, 2^{N+1} \left\|\left(f \circ \cos^{-1}\right)^{(N)}\right\|_{\infty}}{N!}\right) e^{-n^{(1-\alpha)}}\right].$$

Notice in the last the high speed of convergence of order $n^{-\alpha(N+1)}$.

Proof. Call $F := f \circ \cos^{-1}$ and let $\frac{k}{n}, x \in [\varepsilon, \pi - \varepsilon]$. Then

$$f\left(\frac{k}{n}\right) - f(x) = \sum_{j=1}^{N} \frac{F^{(j)}(\cos x)}{j!} \left(\cos \frac{k}{n} - \cos x\right)^j +$$

$$\frac{1}{(N-1)!} \int_x^{\frac{k}{n}} \left(\cos \frac{k}{n} - \cos s\right)^{N-1} \left(F^{(N)}(\cos s) - F^{(N)}(\cos x)\right) d \cos s.$$

Hence

$$\sum_{k=\lceil na \rceil}^{\lfloor nb \rfloor} f\left(\frac{k}{n}\right) \Phi(nx - k) - f(x) \sum_{k=\lceil na \rceil}^{\lfloor nb \rfloor} \Phi(nx - k) =$$

$$\sum_{j=1}^{N} \frac{F^{(j)}(\cos x)}{j!} \sum_{k=\lceil na \rceil}^{\lfloor nb \rfloor} \Phi(nx-k) \left(\cos \frac{k}{n} - \cos x\right)^j + \frac{1}{(N-1)!} \sum_{k=\lceil na \rceil}^{\lfloor nb \rfloor} \Phi(nx-k) \cdot$$

$$\int_x^{\frac{k}{n}} \left(\cos \frac{k}{n} - \cos s\right)^{N-1} \left(F^{(N)}(\cos s) - F^{(N)}(\cos x)\right) d \cos s.$$

Set here $a = \varepsilon$, $b = \pi - \varepsilon$. Thus

$$G_n^*(f, x) - f(x) \sum_{k=\lceil na \rceil}^{\lfloor nb \rfloor} \Phi(nx - k) =$$

$$\sum_{j=1}^{N} \frac{F^{(j)}(\cos x)}{j!} G_n^* \left((\cos \cdot - \cos x)^j, x\right) + \Theta_n(x),$$

where

$$\Theta_n(x) := \sum_{k=\lceil na \rceil}^{\lfloor nb \rfloor} \Phi(nx-k)\,\lambda,$$

with

$$\lambda := \frac{1}{(N-1)!}\int_x^{\frac{k}{n}}\left(\cos\frac{k}{n}-\cos s\right)^{N-1}\left(F^{(N)}(\cos s)-F^{(N)}(\cos x)\right)d\cos s =$$

$$\frac{1}{(N-1)!}\int_{\cos x}^{\cos\frac{k}{n}}\left(\cos\frac{k}{n}-w\right)^{N-1}\left(F^{(N)}(w)-F^{(N)}(\cos x)\right)dw.$$

Case of $\left|\frac{k}{n}-x\right|\le\frac{1}{n^\alpha}$.

i) Subcase of $\frac{k}{n}\ge x$. The function cos ine is decreasing on $[a,b]$, i.e. $\cos\frac{k}{n}\le\cos x$.

Then

$$|\lambda|\le\frac{1}{(N-1)!}\int_{\cos\frac{k}{n}}^{\cos x}\left(w-\cos\frac{k}{n}\right)^{N-1}|\left(F^{(N)}(w)-F^{(N)}(\cos x)\right)|dw\le$$

$$\frac{1}{(N-1)!}\int_{\cos\frac{k}{n}}^{\cos x}\left(w-\cos\frac{k}{n}\right)^{N-1}\omega_1\left(F^{(N)},|w-\cos x|\right)dw\le$$

$$\omega_1\left(F^{(N)},\cos x-\cos\frac{k}{n}\right)\frac{\left(\cos x-\cos\frac{k}{n}\right)^N}{N!}\le$$

$$\omega_1\left(F^{(N)},\left|x-\frac{k}{n}\right|\right)\frac{\left|x-\frac{k}{n}\right|^N}{N!}\le\omega_1\left(F^{(N)},\frac{1}{n^\alpha}\right)\frac{1}{n^{\alpha N}N!}.$$

So if $\frac{k}{n}\ge x$, then

$$|\lambda|\le\omega_1\left(F^{(N)},\frac{1}{n^\alpha}\right)\frac{1}{n^{\alpha N}N!}.$$

ii) Subcase of $\frac{k}{n}\le x$, then $\cos\frac{k}{n}\ge\cos x$. Hence

$$|\lambda|\le\frac{1}{(N-1)!}\int_{\cos x}^{\cos\frac{k}{n}}\left(\cos\frac{k}{n}-w\right)^{N-1}\left|F^{(N)}(w)-F^{(N)}(\cos x)\right|dw\le$$

$$\frac{1}{(N-1)!}\int_{\cos x}^{\cos\frac{k}{n}}\left(\cos\frac{k}{n}-w\right)^{N-1}\omega_1\left(F^{(N)},w-\cos x\right)dw\le$$

$$\frac{1}{(N-1)!}\omega_1\left(F^{(N)},\cos\frac{k}{n}-\cos x\right)\int_{\cos x}^{\cos\frac{k}{n}}\left(\cos\frac{k}{n}-w\right)^{N-1}dw\le$$

$$\frac{1}{(N-1)!}\omega_1\left(F^{(N)},\left|\frac{k}{n}-x\right|\right)\frac{\left(\cos\frac{k}{n}-\cos x\right)^N}{N}\le$$

$$\frac{1}{N!}\omega_1\left(F^{(N)},\left|\frac{k}{n}-x\right|\right)\left|\frac{k}{n}-x\right|^N \le \frac{1}{N!}\omega_1\left(F^{(N)},\frac{1}{n^\alpha}\right)\frac{1}{n^{\alpha N}}.$$

We proved for $\frac{k}{n}\le x$ that

$$|\lambda|\le\omega_1\left(F^{(N)},\frac{1}{n^\alpha}\right)\frac{1}{N!n^{\alpha N}}.$$

So in both cases we got that

$$|\lambda|\le\omega_1\left(F^{(N)},\frac{1}{n^\alpha}\right)\frac{1}{n^{\alpha N}N!},$$

when $\left|\frac{k}{n}-x\right|\le\frac{1}{n^\alpha}$.

Also in general ($\frac{k}{n}\ge x$ case)

$$|\lambda|\le\frac{1}{(N-1)!}\left(\int_{\cos\frac{k}{n}}^{\cos x}\left(w-\cos\frac{k}{n}\right)^{N-1}dw\right)2\left\|F^{(N)}\right\|_\infty\le$$

$$\frac{1}{N!}\left(\cos x-\cos\frac{k}{n}\right)^N 2\left\|F^{(N)}\right\|_\infty\le\frac{2^{N+1}}{N!}\left\|F^{(N)}\right\|_\infty.$$

Also (case of $\frac{k}{n}\le x$) we obtain

$$|\lambda|\le\frac{1}{(N-1)!}\left(\int_{\cos x}^{\cos\frac{k}{n}}\left(\cos\frac{k}{n}-w\right)^{N-1}dw\right)2\left\|F^{(N)}\right\|_\infty=$$

$$\frac{1}{N!}\left(\cos\frac{k}{n}-\cos x\right)^N 2\left\|F^{(N)}\right\|_\infty\le\frac{2^{N+1}}{N!}\left\|F^{(N)}\right\|_\infty.$$

So we proved in general that

$$|\lambda|\le\frac{2^{N+1}}{N!}\left\|F^{(N)}\right\|_\infty.$$

Therefore we derive

$$|\Theta_n(x)|\le\sum_{\substack{k=\lceil na\rceil\\(k:|x-\frac{k}{n}|\le\frac{1}{n^\alpha})}}^{\lfloor nb\rfloor}\Phi(nx-k)|\lambda|+\sum_{\substack{k=\lceil na\rceil\\(k:|x-\frac{k}{n}|>\frac{1}{n^\alpha})}}^{\lfloor nb\rfloor}\Phi(nx-k)|\lambda|\le$$

$$\left(\omega_1\left(F^{(N)},\frac{1}{n^\alpha}\right)\frac{1}{n^{\alpha N}N!}\right)+(3.1992)\frac{2^{N+1}}{N!}\left\|F^{(N)}\right\|_\infty e^{-n^{(1-\alpha)}}.$$

So that

$$|\Theta_n(x)|\le\frac{\omega_1\left(F^{(N)},\frac{1}{n^\alpha}\right)}{n^{\alpha N}N!}+(3.1992)\frac{2^{N+1}}{N!}\left\|F^{(N)}\right\|_\infty e^{-n^{(1-\alpha)}}.$$

Next we estimate

$$\left| G_n^* \left((\cos \cdot - \cos x)^j , x \right) \right| \le G_n^* \left(|\cos \cdot - \cos x|^j , x \right) \le$$

$$G_n^* \left(|\cdot - x|^j , x \right) = \sum_{k=\lceil na \rceil}^{\lfloor nb \rfloor} \Phi (nx - k) \left| \frac{k}{n} - x \right|^j \le$$

(work as before)

$$\frac{1}{n^{\alpha j}} + (3.1992) (\pi - 2\varepsilon)^j e^{-n^{(1-\alpha)}}.$$

Therefore

$$\left| G_n^* \left((\cos \cdot - \cos x)^j , x \right) \right| \le \frac{1}{n^{\alpha j}} + (3.1992) (\pi - 2\varepsilon)^j e^{-n^{(1-\alpha)}},$$

$j = 1, ..., N.$

The theorem is proved.   ■

1.4   Complex Neural Network Quantitative Approximations

We make

Remark 1.13. Let $X := [a, b]$, $\mathbb{R}$ and $f : X \to \mathbb{C}$ with real and imaginary parts $f_1, f_2 : f = f_1 + i f_2$, $i = \sqrt{-1}$. Clearly f is continuous iff f_1 and f_2 are continuous.

Also it holds

$$f^{(j)} (x) = f_1^{(j)} (x) + i f_2^{(j)} (x), \tag{1.40}$$

for all $j = 1, ..., N$, given that $f_1, f_2 \in C^N (X)$, $N \in \mathbb{N}$.

We denote by $C_B (\mathbb{R}, \mathbb{C})$ the space of continuous and bounded functions $f : \mathbb{R} \to \mathbb{C}$. Clearly f is bounded, iff both f_1, f_2 are bounded from $\mathbb{R}$ into $\mathbb{R}$, where $f = f_1 + i f_2$.

Here we define

$$G_n (f, x) := G_n (f_1, x) + i G_n (f_2, x), \tag{1.41}$$

and

$$\overline{G}_n (f, x) := \overline{G}_n (f_1, x) + i \overline{G}_n (f_2, x). \tag{1.42}$$

We observe here that

$$|G_n (f, x) - f (x)| \le |G_n (f_1, x) - f_1 (x)| + |G_n (f_2, x) - f_2 (x)|, \tag{1.43}$$

and

$$\left\| G_n \left(f \right) - f \right\|_\infty \le \left\| G_n \left(f_1 \right) - f_1 \right\|_\infty + \left\| G_n \left(f_2 \right) - f_2 \right\|_\infty. \tag{1.44}$$

Similarly we get

$$\left| \overline{G}_n \left(f, x \right) - f \left(x \right) \right| \le \left| \overline{G}_n \left(f_1, x \right) - f_1 \left(x \right) \right| + \left| \overline{G}_n \left(f_2, x \right) - f_2 \left(x \right) \right|, \tag{1.45}$$

and

$$\left\| \overline{G}_n \left(f \right) - f \right\|_\infty \le \left\| \overline{G}_n \left(f_1 \right) - f_1 \right\|_\infty + \left\| \overline{G}_n \left(f_2 \right) - f_2 \right\|_\infty. \tag{1.46}$$

We give

Theorem 1.14. *Let* $f \in C \left([a, b], \mathbb{C} \right)$, $f = f_1 + i f_2$, $0 < \alpha < 1$, $n \in \mathbb{N}$, $x \in [a, b]$. *Then*
  i)

$$\left| G_n \left(f, x \right) - f \left(x \right) \right| \le \left(5.250312578 \right) \cdot \tag{1.47}$$

$$\left[\left(\omega_1 \left(f_1, \frac{1}{n^\alpha} \right) + \omega_1 \left(f_2, \frac{1}{n^\alpha} \right) \right) + \left(6.3984 \right) \left(\left\| f_1 \right\|_\infty + \left\| f_2 \right\|_\infty \right) e^{-n^{(1-\alpha)}} \right] =: \psi_1,$$

and
  ii)

$$\left\| G_n \left(f \right) - f \right\|_\infty \le \psi_1. \tag{1.48}$$

Proof. Based on Remark 1.13 and Theorem 1.6. ∎

We give

Theorem 1.15. *Let* $f \in C_B \left(\mathbb{R}, \mathbb{C} \right)$, $f = f_1 + i f_2$, $0 < \alpha < 1$, $n \in \mathbb{N}$, $x \in \mathbb{R}$. *Then*
  i)

$$\left| \overline{G}_n \left(f, x \right) - f \left(x \right) \right| \le \left(\omega_1 \left(f_1, \frac{1}{n^\alpha} \right) + \omega_1 \left(f_2, \frac{1}{n^\alpha} \right) \right) + \tag{1.49}$$

$$\left(6.3984 \right) \left(\left\| f_1 \right\|_\infty + \left\| f_2 \right\|_\infty \right) e^{-n^{(1-\alpha)}} =: \psi_2,$$

ii)

$$\left\| \overline{G}_n \left(f \right) - f \right\|_\infty \le \psi_2. \tag{1.50}$$

Proof. Based on Remark 1.13 and Theorem 1.7. ∎

Next we present a result of high order complex neural network approximation.

Theorem 1.16. *Let* $f : [a, b] \to \mathbb{C}$, $[a, b] \subset \mathbb{R}$, *such that* $f = f_1 + i f_2$. *Assume* $f_1, f_2 \in C^N \left([a, b] \right)$, $n, N \in \mathbb{N}$, $0 < \alpha < 1$, $x \in [a, b]$. *Then*
  i)

$$\left| G_n \left(f, x \right) - f \left(x \right) \right| \le \left(5.250312578 \right) \cdot \tag{1.51}$$

$$\left\{ \sum_{j=1}^{N} \frac{\left(\left| f_1^{(j)}(x) \right| + \left| f_2^{(j)}(x) \right| \right)}{j!} \left[\frac{1}{n^{\alpha j}} + (3.1992)(b-a)^j \, e^{-n^{(1-\alpha)}} \right] + \right.$$

$$\left[\frac{\left(\omega_1 \left(f_1^{(N)}, \frac{1}{n^\alpha} \right) + \omega_1 \left(f_2^{(N)}, \frac{1}{n^\alpha} \right) \right)}{n^{\alpha N} N!} + \right.$$

$$\left. \left. \left(\frac{(6.3984) \left(\left\| f_1^{(N)} \right\|_\infty + \left\| f_2^{(N)} \right\|_\infty \right) (b-a)^N}{N!} \right) e^{-n^{(1-\alpha)}} \right] \right\},$$

ii) assume further $f_1^{(j)}(x_0) = f_2^{(j)}(x_0) = 0$, $j = 1, ..., N$, *for some* $x_0 \in [a, b]$, *it holds*

$$|G_n(f, x_0) - f(x_0)| \leq (5.250312578) \cdot \qquad (1.52)$$

$$\left[\frac{\left(\omega_1 \left(f_1^{(N)}, \frac{1}{n^\alpha} \right) + \omega_1 \left(f_2^{(N)}, \frac{1}{n^\alpha} \right) \right)}{n^{\alpha N} N!} + \right.$$

$$\left. \left(\frac{(6.3984) \left(\left\| f_1^{(N)} \right\|_\infty + \left\| f_2^{(N)} \right\|_\infty \right) (b-a)^N}{N!} \right) e^{-n^{(1-\alpha)}} \right],$$

notice here the extremely high rate of convergence at $n^{-(N+1)\alpha}$,
iii)

$$\|G_n(f) - f\|_\infty \leq (5.250312578) \cdot$$

$$\left\{ \sum_{j=1}^{N} \frac{\left(\left\| f_1^{(j)} \right\|_\infty + \left\| f_2^{(j)} \right\|_\infty \right)}{j!} \left[\frac{1}{n^{\alpha j}} + (3.1992)(b-a)^j \, e^{-n^{(1-\alpha)}} \right] + \right.$$

$$\left[\frac{\left(\omega_1 \left(f_1^{(N)}, \frac{1}{n^\alpha} \right) + \omega_1 \left(f_2^{(N)}, \frac{1}{n^\alpha} \right) \right)}{n^{\alpha N} N!} + \right.$$

$$\left. \left. \left(\frac{(6.3984) \left(\left\| f_1^{(N)} \right\|_\infty + \left\| f_2^{(N)} \right\|_\infty \right) (b-a)^N}{N!} \right) e^{-n^{(1-\alpha)}} \right] \right\}. \qquad (1.53)$$

Proof. Based on Remark 1.13 and Theorem 1.8.   ∎

References

[1] Anastassiou, G.A.: Rate of convergence of some neural network operators to the unit-univariate case. J. Math. Anal. Appli. 212, 237–262 (1997)

[2] Anastassiou, G.A.: Quantitative Approximations. Chapman&Hall/CRC, Boca Raton (2001)

[3] Anastassiou, G.A.: Basic Inequalities, Revisited. Mathematica Balkanica, New Series 24, Fasc. 1-2, 59–84 (2010)

[4] Anastassiou, G.A.: Univariate sigmoidal neural network approximation. Journal of Comp. Anal. and Appl. (accepted 2011)

[5] Barron, A.R.: Universal approximation bounds for superpositions of a sigmoidal function. IEEE Trans. Inform. Theory 39, 930–945 (1993)

[6] Brauer, F., Castillo-Chavez, C.: Mathematical models in population biology and epidemiology, pp. 8–9. Springer, New York (2001)

[7] Cao, F.L., Xie, T.F., Xu, Z.B.: The estimate for approximation error of neural networks: a constructive approach. Neurocomputing 71, 626–630 (2008)

[8] Chen, T.P., Chen, H.: Universal approximation to nonlinear operators by neural networks with arbitrary activation functions and its applications to a dynamic system. IEEE Trans. Neural Networks 6, 911–917 (1995)

[9] Chen, Z., Cao, F.: The approximation operators with sigmoidal functions. Computers and Mathematics with Applications 58, 758–765 (2009)

[10] Chui, C.K., Li, X.: Approximation by ridge functions and neural networks with one hidden layer. J. Approx. Theory 70, 131–141 (1992)

[11] Cybenko, G.: Approximation by superpositions of sigmoidal function. Math. of Control Signals and System 2, 303–314 (1989)

[12] Ferrari, S., Stengel, R.F.: Smooth function approximation using neural networks. IEEE Trans. Neural Networks 16, 24–38 (2005)

[13] Funahashi, K.I.: On the approximate realization of continuous mappings by neural networks. Neural Networks 2, 183–192 (1989)

[14] Hahm, N., Hong, B.I.: An approximation by neural networks with a fixed weight. Computers & Math. with Appli. 47, 1897–1903 (2004)

[15] Hornik, K., Stinchombe, M., White, H.: Multilayer feedforward networks are universal approximators. Neural Networks 2, 359–366 (1989)

[16] Hornik, K., Stinchombe, M., White, H.: Universal approximation of an unknown mapping and its derivatives using multilayer feedforward networks. Neural Networks 3, 551–560 (1990)

[17] Hritonenko, N., Yatsenko, Y.: Mathematical modeling in economics, ecology and the environment, pp. 92–93. Science Press, Beijing (2006) (reprint)

[18] Leshno, M., Lin, V.Y., Pinks, A., Schocken, S.: Multilayer feedforward networks with a nonpolynomial activation function can approximate any function. Neural Networks 6, 861–867 (1993)

[19] Lorentz, G.G.: Approximation of Functions. Rinehart and Winston, New York (1966)

[20] Maiorov, V., Meir, R.S.: Approximation bounds for smooth functions in $C(R^d)$ by neural and mixture networks. IEEE Trans. Neural Networks 9, 969–978 (1998)

[21] Makovoz, Y.: Uniform approximation by neural networks. J. Approx. Theory 95, 215–228 (1998)

[22] Mhaskar, H.N., Micchelli, C.A.: Approximation by superposition of a sigmoidal function. Adv. Applied Math. 13, 350–373 (1992)

[23] Mhaskar, H.N., Micchelli, C.A.: Degree of approximation by neural networks with a single hidden layer. Adv. Applied Math. 16, 151–183 (1995)

[24] Suzuki, S.: Constructive function approximation by three-layer artificial neural networks. Neural Networks 11, 1049–1058 (1998)

[25] Xie, T.F., Zhou, S.P.: Approximation Theory of Real Functions. Hangzhou University Press, Hangzhou (1998)

[26] Xu, Z.B., Cao, F.L.: The essential order of approximation for neural networks. Science in China (Ser. F) 47, 97–112 (2004)

Chapter 2
Univariate Hyperbolic Tangent Neural Network Quantitative Approximation

Here we give the univariate quantitative approximation of real and complex valued continuous functions on a compact interval or all the real line by quasi-interpolation hyperbolic tangent neural network operators. This approximation is obtained by establishing Jackson type inequalities involving the modulus of continuity of the engaged function or its high order derivative. The operators are defined by using a density function induced by the hyperbolic tangent function. Our approximations are pointwise and with respect to the uniform norm. The related feed-forward neural network is with one hidden layer. This chapter relies on [4].

2.1 Introduction

The author in [1] and [2], see chapters 2-5, was the first to present neural network approximations to continuous functions with rates by very specifically defined neural network operators of Cardaliagnet-Euvrard and "Squashing" types, by employing the modulus of continuity of the engaged function or its high order derivative, and producing very tight Jackson type inequalities. He treats there both the univariate and multivariate cases. The defining these operators "bell-shaped" and "squashing" functions are assumed to be of compact support. Also in [2] he gives the Nth order asymptotic expansion for the error of weak approximation of these two operators to a special natural class of smooth functions, see chapters 4-5 there.

For this chapter the author is inspired by the article [5] by Z. Chen and F. Cao.

He does related to it work and much more beyond. So the author here gives univariate hyperbolic tangent neural network approximations to continuous functions over compact intervals of the real line or over the whole $\mathbb{R}$, then he extends his results to complex valued functions. All convergences here are with rates expressed via the modulus of continuity of the involved function or its high order derivative, and given by very tight Jackson type inequalities.

The author here comes up with the "right" precisely defined quasi-interpolation neural network operator, associated with hyperbolic tangent function and related to a compact interval or real line. The compact intervals are not necessarily symmetric to the origin. Some of the upper bounds to error quantity are very flexible and general. In preparation to establish our results we give important properties of the basic density function defining our operators.

Feed-forward neural networks (FNNs) with one hidden layer, the only type of networks we deal with in this chapter, are mathematically expressed as

$$N_n(x) = \sum_{j=0}^{n} c_j \sigma\left(\langle a_j \cdot x\rangle + b_j\right), \quad x \in \mathbb{R}^s, \quad s \in \mathbb{N},$$

where for $0 \leq j \leq n$, $b_j \in \mathbb{R}$ are the thresholds, $a_j \in \mathbb{R}^s$ are the connection weights, $c_j \in \mathbb{R}$ are the coefficients, $\langle a_j \cdot x\rangle$ is the inner product of a_j and x, and σ is the activation function of the network. In many fundamental network models, the activation function is the hyperbolic tangent. About neural networks see [6], [7], [8].

2.2 Basic Ideas

We consider here the hyperbolic tangent function $\tanh x$, $x \in \mathbb{R}$:

$$\tanh x := \frac{e^x - e^{-x}}{e^x + e^{-x}} = \frac{e^{2x} - 1}{e^{2x} + 1}.$$

It has the properties $\tanh 0 = 0$, $-1 < \tanh x < 1$, $\forall\, x \in \mathbb{R}$, and $\tanh(-x) = -\tanh x$. Furthermore $\tanh x \to 1$ as $x \to \infty$, and $\tanh x \to -1$, as $x \to -\infty$, and it is strictly increasing on $\mathbb{R}$. Furthermore it holds $\frac{d}{dx}\tanh x = \frac{1}{\cosh^2 x} > 0$.

This function plays the role of an activation function in the hidden layer of neural networks.

We further consider

$$\Psi(x) := \frac{1}{4}\left(\tanh(x+1) - \tanh(x-1)\right) > 0, \quad \forall\, x \in \mathbb{R}.$$

We easily see that $\Psi(-x) = \Psi(x)$, that is Ψ is even on $\mathbb{R}$. Obviously Ψ is differentiable, thus continuous.

Proposition 2.1. $\Psi(x)$ *for* $x \geq 0$ *is strictly decreasing.*

Proof. Clearly

$$\frac{d\Psi(x)}{dx} = \frac{1}{4}\left(\frac{1}{\cosh^2(x+1)} - \frac{1}{\cosh^2(x-1)}\right)$$

$$= \frac{\left(\cosh^2(x-1) - \cosh^2(x+1)\right)}{4\cosh^2(x+1)\cosh^2(x-1)}$$

$$= \left(\frac{\cosh(x-1) + \cosh(x+1)}{4\cosh^2(x+1)\cosh^2(x-1)}\right)\left(\cosh(x-1) - \cosh(x+1)\right).$$

Suppose $x - 1 \geq 0$. Clearly $x - 1 < x + 1$, so that $0 < \cosh(x-1) < \cosh(x+1)$ and $\left(\cosh(x-1) - \cosh(x+1)\right) < 0$, hence $\Psi'(x) < 0$.

Let now $x - 1 < 0$, then $1 - x > 0$, and for $x > 0$ we have $0 < \cosh(x-1) = \cosh(1-x) < \cosh(1+x) = \cosh(x+1)$. Hence $\left(\cosh(x-1) - \cosh(x+1)\right) < 0$, proving again $\Psi'(x) < 0$. Also Ψ is continuous everywhere and in particular at zero.

The claim is proved.   ∎

Clearly $\Psi(x)$ is strictly increasing for $x \leq 0$. Also it holds $\lim\limits_{x \to -\infty} \Psi(x) = 0 = \lim\limits_{x \to \infty} \Psi(x)$.

Infact Ψ has the bell shape with horizontal asymptote the x-axis. So the maximum of Ψ is zero, $\Psi(0) = 0.3809297$.

Theorem 2.2. *We have that* $\sum_{i=-\infty}^{\infty} \Psi(x-i) = 1$, $\forall\, x \in \mathbb{R}$.

Proof. We observe that

$$\sum_{i=-\infty}^{\infty} \left(\tanh(x-i) - \tanh(x-1-i)\right) =$$

$$\sum_{i=0}^{\infty} \left(\tanh(x-i) - \tanh(x-1-i)\right) + \sum_{i=-\infty}^{-1} \left(\tanh(x-i) - \tanh(x-1-i)\right).$$

Furthermore $(\lambda \in \mathbb{Z}^+)$

$$\sum_{i=0}^{\infty} \left(\tanh(x-i) - \tanh(x-1-i)\right) =$$

$$\lim_{\lambda \to \infty} \sum_{i=0}^{\lambda} \left(\tanh(x-i) - \tanh(x-1-i)\right) =$$

(telescoping sum)

$$\lim_{\lambda \to \infty} (\tanh x - \tanh (x - \lambda)) = 1 + \tanh x.$$

Similarly,

$$\sum_{i=-\infty}^{-1} (\tanh (x - i) - \tanh (x - 1 - i)) =$$

$$\lim_{\lambda \to \infty} \sum_{i=-\lambda}^{-1} (\tanh (x - i) - \tanh (x - 1 - i)) =$$

$$\lim_{\lambda \to \infty} (\tanh (x + \lambda) - \tanh x) = 1 - \tanh x.$$

So adding the last two limits we obtain

$$\sum_{i=-\infty}^{\infty} (\tanh (x - i) - \tanh (x - 1 - i)) = 2, \quad \forall\, x \in \mathbb{R}.$$

Similarly we obtain

$$\sum_{i=-\infty}^{\infty} (\tanh (x + 1 - i) - \tanh (x - i)) = 2, \quad \forall\, x \in \mathbb{R}.$$

Consequently

$$\sum_{i=-\infty}^{\infty} (\tanh (x + 1 - i) - \tanh (x - 1 - i)) = 4, \quad \forall\, x \in \mathbb{R},$$

proving the claim.   ∎

Thus

$$\sum_{i=-\infty}^{\infty} \Psi (nx - i) = 1, \quad \forall\, n \in \mathbb{N},\ \forall\, x \in \mathbb{R}.$$

Furthermore we give:

Because Ψ is even it holds $\sum_{i=-\infty}^{\infty} \Psi (i - x) = 1$, $\forall x \in \mathbb{R}$.

Hence $\sum_{i=-\infty}^{\infty} \Psi (i + x) = 1$, $\forall\, x \in \mathbb{R}$, and $\sum_{i=-\infty}^{\infty} \Psi (x + i) = 1$, $\forall\, x \in \mathbb{R}$.

Theorem 2.3. *It holds $\int_{-\infty}^{\infty} \Psi (x)\, dx = 1$.*

Proof. We observe that

$$\int_{-\infty}^{\infty} \Psi (x)\, dx = \sum_{j=-\infty}^{\infty} \int_{j}^{j+1} \Psi (x)\, dx = \sum_{j=-\infty}^{\infty} \int_{0}^{1} \Psi (x + j)\, dx =$$

$$\int_0^1 \left(\sum_{j=-\infty}^{\infty} \Psi(x+j) \right) dx = \int_0^1 1 dx = 1.$$ ∎

So $\Psi(x)$ is a density function on $\mathbb{R}$.

Theorem 2.4. *Let $0 < \alpha < 1$ and $n \in \mathbb{N}$. It holds*

$$\sum_{\substack{k=-\infty \\ : |nx-k| \geq n^{1-\alpha}}}^{\infty} \Psi(nx-k) \leq e^4 \cdot e^{-2n^{(1-\alpha)}}.$$

Proof. Let $x \geq 1$. That is $0 \leq x - 1 < x + 1$.
  Applying the mean value theorem we obtain

$$\Psi(x) = \frac{1}{4} \cdot 2 \cdot \frac{1}{\cosh^2 \xi} = \frac{1}{2} \frac{1}{\cosh^2 \xi},$$

for some $x - 1 < \xi < x + 1$.
  We get $\cosh(x-1) < \cosh \xi < \cosh(x+1)$ and $\cosh^2(x-1) < \cosh^2 \xi < \cosh^2(x+1)$.
  Therefore

$$0 < \frac{1}{\cosh^2 \xi} < \frac{1}{\cosh^2(x-1)},$$

and

$$\Psi(x) < \frac{1}{2} \frac{1}{\cosh^2(x-1)} = \frac{1}{2} \frac{4}{(e^{x-1} + e^{1-x})^2} = \frac{2}{(e^{x-1} + e^{1-x})^2} =: (*).$$

From $e^{x-1} + e^{1-x} > e^{x-1}$ we obtain $(e^{x-1} + e^{1-x})^2 > e^{2(x-1)}$, and

$$\frac{1}{(e^{x-1} + e^{1-x})^2} < \frac{1}{e^{2(x-1)}}.$$

So that

$$(*) < \frac{2}{e^{2(x-1)}} = 2e^2 \cdot e^{-2x}.$$

That is we proved

$$\Psi(x) < 2e^2 \cdot e^{-2x}, \quad x \geq 1.$$

Thus

$$\sum_{\substack{k=-\infty \\ : |nx-k| \geq n^{1-\alpha}}}^{\infty} \Psi(nx-k) = \sum_{\substack{k=-\infty \\ : |nx-k| \geq n^{1-\alpha}}}^{\infty} \Psi(|nx-k|) <$$

$$2e^2 \sum_{\substack{k=-\infty \\ \{\,:\,|nx-k| \geq n^{1-\alpha}}} \left(e^{-2|nx-k|}\right) \leq 2e^2 \int_{(n^{1-\alpha}-1)}^{\infty} e^{-2x}\,dx$$

$$= e^2 \int_{2(n^{1-\alpha}-1)}^{\infty} e^{-y}\,dy = e^2 e^{-2\left(n^{(1-\alpha)}-1\right)} = e^4 \cdot e^{-2n^{(1-\alpha)}},$$

proving the claim.   ∎

Denote by $\lfloor \cdot \rfloor$ the integral part of the number and by $\lceil \cdot \rceil$ the ceiling of the number.

We further give

Theorem 2.5. *Let $x \in [a, b] \subset \mathbb{R}$ and $n \in \mathbb{N}$ so that $\lceil na \rceil \leq \lfloor nb \rfloor$. It holds*

$$\frac{1}{\sum_{k=\lceil na \rceil}^{\lfloor nb \rfloor} \Psi\left(nx - k\right)} < 4.1488766 = \frac{1}{\Psi\left(1\right)}.$$

Proof. We observe that

$$1 = \sum_{k=-\infty}^{\infty} \Psi\left(nx - k\right) > \sum_{k=\lceil na \rceil}^{\lfloor nb \rfloor} \Psi\left(nx - k\right) =$$

$$\sum_{k=\lceil na \rceil}^{\lfloor nb \rfloor} \Psi\left(|nx - k|\right) > \Psi\left(|nx - k_0|\right),$$

$\forall\, k_0 \in [\lceil na \rceil, \lfloor nb \rfloor] \cap \mathbb{Z}$.

We can choose $k_0 \in [\lceil na \rceil, \lfloor nb \rfloor] \cap \mathbb{Z}$ such that $|nx - k_0| < 1$.
Therefore

$$\Psi\left(|nx - k_0|\right) > \Psi\left(1\right) = \frac{1}{4}\left(\tanh\left(2\right) - \tanh\left(0\right)\right) =$$

$$\frac{1}{4}\tanh 2 = \frac{1}{4}\left(\frac{e^2 - e^{-2}}{e^2 + e^{-2}}\right) = \frac{1}{4}\left(\frac{e^4 - 1}{e^4 + 1}\right) = 0.2410291.$$

So $\Psi\left(1\right) = 0.2410291$.

Consequently we obtain

$$\sum_{k=\lceil na \rceil}^{\lfloor nb \rfloor} \Psi\left(|nx - k|\right) > 0.2410291 = \Psi\left(1\right),$$

and

$$\frac{1}{\sum_{k=\lceil na \rceil}^{\lfloor nb \rfloor} \Psi\left(|nx - k|\right)} < 4.1488766 = \frac{1}{\Psi\left(1\right)},$$

proving the claim.   ∎

Remark 2.6. *We also notice that*

$$1 - \sum_{k=\lceil na\rceil}^{\lfloor nb\rfloor} \Psi(nb-k) = \sum_{k=-\infty}^{\lceil na\rceil-1} \Psi(nb-k) + \sum_{k=\lfloor nb\rfloor+1}^{\infty} \Psi(nb-k)$$

$$> \Psi(nb - \lfloor nb\rfloor - 1)$$

(call $\varepsilon := nb - \lfloor nb\rfloor$, $0 \le \varepsilon < 1$)

$$= \Psi(\varepsilon - 1) = \Psi(1-\varepsilon) \ge \Psi(1) > 0.$$

Therefore $\lim\limits_{n\to\infty} \left(1 - \sum_{k=\lceil na\rceil}^{\lfloor nb\rfloor} \Psi(nb-k)\right) > 0.$
Similarly,

$$1 - \sum_{k=\lceil na\rceil}^{\lfloor nb\rfloor} \Psi(na-k) = \sum_{k=-\infty}^{\lceil na\rceil-1} \Psi(na-k) + \sum_{k=\lfloor nb\rfloor+1}^{\infty} \Psi(na-k)$$

$$> \Psi(na - \lceil na\rceil + 1)$$

(call $\eta := \lceil na\rceil - na$, $0 \le \eta < 1$)

$$= \Psi(1-\eta) \ge \Psi(1) > 0.$$

Therefore again $\lim\limits_{n\to\infty} \left(1 - \sum_{k=\lceil na\rceil}^{\lfloor nb\rfloor} \Psi(na-k)\right) > 0.$
Therefore we find that

$$\lim_{n\to\infty} \sum_{k=\lceil na\rceil}^{\lfloor nb\rfloor} \Psi(nx-k) \ne 1,$$

for at least some $x \in [a,b]$.

Definition 2.7. Let $f \in C([a,b])$ and $n \in \mathbb{N}$ such that $\lceil na\rceil \le \lfloor nb\rfloor$.
We introduce and define the positive linear neural network operator

$$F_n(f,x) := \frac{\sum_{k=\lceil na\rceil}^{\lfloor nb\rfloor} f\left(\frac{k}{n}\right) \Psi(nx-k)}{\sum_{k=\lceil na\rceil}^{\lfloor nb\rfloor} \Psi(nx-k)}, \quad x \in [a.b]. \tag{2.1}$$

For large enough n we always obtain $\lceil na\rceil \le \lfloor nb\rfloor$. Also $a \le \frac{k}{n} \le b$, iff $\lceil na\rceil \le k \le \lfloor nb\rfloor$.

We study here the pointwise and uniform convergence of $F_n(f,x)$ to $f(x)$ with rates.

For convenience we call

$$F_n^*(f,x) := \sum_{k=\lceil na\rceil}^{\lfloor nb\rfloor} f\left(\frac{k}{n}\right) \Psi(nx-k), \tag{2.2}$$

that is

$$F_n\left(f,x\right) := \frac{F_n^*\left(f,x\right)}{\sum_{k=\lceil na\rceil}^{\lfloor nb\rfloor} \Psi\left(nx-k\right)}. \qquad (2.3)$$

So that,

$$F_n\left(f,x\right) - f\left(x\right) = \frac{F_n^*\left(f,x\right)}{\sum_{k=\lceil na\rceil}^{\lfloor nb\rfloor} \Psi\left(nx-k\right)} - f\left(x\right)$$

$$= \frac{F_n^*\left(f,x\right) - f\left(x\right)\sum_{k=\lceil na\rceil}^{\lfloor nb\rfloor} \Psi\left(nx-k\right)}{\sum_{k=\lceil na\rceil}^{\lfloor nb\rfloor} \Psi\left(nx-k\right)}. \qquad (2.4)$$

Consequently we derive

$$\left|F_n\left(f,x\right) - f\left(x\right)\right| \le \frac{1}{\Psi\left(1\right)}\left|F_n^*\left(f,x\right) - f\left(x\right)\sum_{k=\lceil na\rceil}^{\lfloor nb\rfloor} \Psi\left(nx-k\right)\right|. \qquad (2.5)$$

That is

$$\left|F_n\left(f,x\right) - f\left(x\right)\right| \le \left(4.1488766\right)\left|\sum_{k=\lceil na\rceil}^{\lfloor nb\rfloor} \left(f\left(\frac{k}{n}\right) - f\left(x\right)\right)\Psi\left(nx-k\right)\right|. \qquad (2.6)$$

We will estimate the right hand side of (2.6).

For that we need, for $f \in C\left(\left[a,b\right]\right)$ the first modulus of continuity

$$\omega_1\left(f,h\right) := \sup_{\substack{x,y \in [a,b] \\ |x-y| \le h}} \left|f\left(x\right) - f\left(y\right)\right|, \quad h > 0. \qquad (2.7)$$

Similarly it is defined for $f \in C_B\left(\mathbb{R}\right)$ (continuous and bounded on $\mathbb{R}$). We have that $\lim_{h\to 0}\omega_1\left(f,h\right) = 0$.

Definition 2.8. When $f \in C_B\left(\mathbb{R}\right)$ we define,

$$\overline{F}_n\left(f,x\right) := \sum_{k=-\infty}^{\infty} f\left(\frac{k}{n}\right)\Psi\left(nx-k\right), \quad n \in \mathbb{N}, x \in \mathbb{R}, \qquad (2.8)$$

the quasi-interpolation neural network operator.

By [3] we derive the following three theorems on extended Taylor formula.

Theorem 2.9. Let $N \in \mathbb{N}$, $0 < \varepsilon < \frac{\pi}{2}$ small, and $f \in C^N\left(\left[-\frac{\pi}{2}+\varepsilon, \frac{\pi}{2}-\varepsilon\right]\right)$; $x,y \in \left[-\frac{\pi}{2}+\varepsilon, \frac{\pi}{2}-\varepsilon\right]$. Then

$$f\left(x\right) = f\left(y\right) + \sum_{k=1}^{N} \frac{\left(f\circ\sin^{-1}\right)^{(k)}\left(\sin y\right)}{k!}\left(\sin x - \sin y\right)^k + K_N\left(y,x\right), \qquad (2.9)$$

where

$$K_N(y,x) = \frac{1}{(N-1)!}.$$ (2.10)

$$\int_y^x (\sin x - \sin s)^{N-1} \left(\left(f \circ \sin^{-1} \right)^{(N)} (\sin s) - \left(f \circ \sin^{-1} \right)^{(N)} (\sin y) \right) \cos s \, ds.$$

Theorem 2.10. *Let $f \in C^N([\varepsilon, \pi - \varepsilon])$, $N \in \mathbb{N}$, $\varepsilon > 0$ small; $x, y \in [\varepsilon, \pi - \varepsilon]$.*
Then

$$f(x) = f(y) + \sum_{k=1}^N \frac{\left(f \circ \cos^{-1} \right)^{(k)} (\cos y)}{k!} (\cos x - \cos y)^k + K_N^*(y,x), \quad (2.11)$$

where

$$K_N^*(y,x) = -\frac{1}{(N-1)!}.$$ (2.12)

$$\int_y^x (\cos x - \cos s)^{N-1} \left[\left(f \circ \cos^{-1} \right)^{(N)} (\cos s) - \left(f \circ \cos^{-1} \right)^{(N)} (\cos y) \right] \sin s \, ds.$$

Theorem 2.11. *Let $f \in C^N([a,b])$ (or $f \in C^N(\mathbb{R})$), $N \in \mathbb{N}$; $x, y \in [a,b]$*
(or $x, y \in \mathbb{R}$). Then

$$f(x) = f(y) + \sum_{k=1}^N \frac{\left(f \circ \ln \frac{1}{e} \right)^{(k)} (e^{-y})}{k!} (e^{-x} - e^{-y})^k + \overline{K}_N(y,x), \quad (2.13)$$

where

$$\overline{K}_N(y,x) = -\frac{1}{(N-1)!}.$$ (2.14)

$$\int_y^x (e^{-x} - e^{-s})^{N-1} \left[\left(f \circ \ln \frac{1}{e} \right)^{(N)} (e^{-s}) - \left(f \circ \ln \frac{1}{e} \right)^{(N)} (e^{-y}) \right] e^{-s} ds.$$

Remark 2.12. *Using the mean value theorem we obtain*

$$|\sin x - \sin y| \le |x - y|,$$ (2.15)
$$|\cos x - \cos y| \le |x - y|, \quad \forall \, x, y \in \mathbb{R},$$

furthermore we have

$$|\sin x - \sin y| \le 2,$$ (2.16)
$$|\cos x - \cos y| \le 2, \quad \forall \, x, y \in \mathbb{R}.$$

Similarly we get

$$\left| e^{-x} - e^{-y} \right| \le e^{-a} |x - y|,$$ (2.17)

and

$$\left| e^{-x} - e^{-y} \right| \le e^{-a} - e^{-b}, \quad \forall \, x, y \in [a, b].$$ (2.18)

Let $g(x) = \ln_{\frac{1}{e}} x$, $\sin^{-1} x$, $\cos^{-1} x$ and assume $f^{(j)}(x_0) = 0$, $k = 1, ..., N$. Then, by [3], we get $\left(f \circ g^{-1}\right)^{(j)} (g(x_0)) = 0$, $j = 1, ..., N$.

Remark 2.13. *It is well known that* $e^x > x^m$, $m \in \mathbb{N}$, *for large* $x > 0$. *Let fixed* $\alpha, \beta > 0$, *then* $\left\lceil \frac{\alpha}{\beta} \right\rceil \in \mathbb{N}$, *and for large* $x > 0$ *we have*

$$e^x > x^{\left\lceil \frac{\alpha}{\beta} \right\rceil} \geq x^{\frac{\alpha}{\beta}}.$$

So for suitable very large $x > 0$ *we find*

$$e^{x^\beta} > \left(x^\beta\right)^{\frac{\alpha}{\beta}} = x^\alpha.$$

We proved for large $x > 0$ *and* $\alpha, \beta > 0$ *that*

$$e^{x^\beta} > x^\alpha. \tag{2.19}$$

Therefore for large $n \in \mathbb{N}$ *and fixed* $\alpha, \beta > 0$, *we have*

$$e^{2n^\beta} > n^\alpha. \tag{2.20}$$

That is

$$e^{-2n^\beta} < n^{-\alpha}, \quad \text{for large } n \in \mathbb{N}. \tag{2.21}$$

So for $0 < \alpha < 1$ *we get*

$$e^{-2n^{(1-\alpha)}} < n^{-\alpha}. \tag{2.22}$$

Thus be given fixed $A, B > 0$, for the linear combination $\left(An^{-\alpha} + Be^{-2n^{(1-\alpha)}}\right)$ the (dominant) rate of convergence to zero is $n^{-\alpha}$.

The closer α is to 1 we get faster and better rate of convergence to zero.

2.3   Real Neural Network Quantitative Approximations

Here we give a series of neural network approximation to a function given with rates.

We first present

Theorem 2.14. *Let* $f \in C([a, b])$, $0 < \alpha < 1$, $n \in \mathbb{N}$, $x \in [a, b]$. *Then*
   i)

$$|F_n(f, x) - f(x)| \leq (4.1488766) \left[\omega_1\left(f, \frac{1}{n^\alpha}\right) + 2e^4 \|f\|_\infty e^{-2n^{(1-\alpha)}}\right] =: \lambda^*, \tag{2.23}$$

and
   ii)

$$\|F_n(f) - f\|_\infty \leq \lambda^*, \tag{2.24}$$

where $\|\cdot\|_\infty$ *is the supremum norm.*

Proof. We see that

$$\left| \sum_{k=\lceil na \rceil}^{\lfloor nb \rfloor} \left(f\left(\frac{k}{n}\right) - f(x) \right) \Psi(nx - k) \right| \le$$

$$\sum_{k=\lceil na \rceil}^{\lfloor nb \rfloor} \left| f\left(\frac{k}{n}\right) - f(x) \right| \Psi(nx - k) =$$

$$\sum_{\substack{k = \lceil na \rceil \\ \left| \frac{k}{n} - x \right| \le \frac{1}{n^\alpha}}}^{\lfloor nb \rfloor} \left| f\left(\frac{k}{n}\right) - f(x) \right| \Psi(nx - k) +$$

$$\sum_{\substack{k = \lceil na \rceil \\ \left| \frac{k}{n} - x \right| > \frac{1}{n^\alpha}}}^{\lfloor nb \rfloor} \left| f\left(\frac{k}{n}\right) - f(x) \right| \Psi(nx - k) \le$$

$$\sum_{\substack{k = \lceil na \rceil \\ \left| \frac{k}{n} - x \right| \le \frac{1}{n^\alpha}}}^{\lfloor nb \rfloor} \omega_1\left(f, \left| \frac{k}{n} - x \right| \right) \Psi(nx - k) +$$

$$2\|f\|_\infty \sum_{\substack{k = \lceil na \rceil \\ |k - nx| > n^{1-\alpha}}}^{\lfloor nb \rfloor} \Psi(nx - k) \le$$

$$\omega_1\left(f, \frac{1}{n^\alpha} \right) \sum_{\substack{k = -\infty \\ \left| \frac{k}{n} - x \right| \le \frac{1}{n^\alpha}}}^{\infty} \Psi(nx - k) +$$

$$2\|f\|_\infty \sum_{\substack{k = -\infty \\ |k - nx| > n^{1-\alpha}}}^{\infty} \Psi(nx - k) \underset{\text{(by Theorem 2.4)}}{\le}$$

$$\omega_1\left(f, \frac{1}{n^\alpha} \right) + 2\|f\|_\infty\, e^4 e^{-2n^{(1-\alpha)}}$$

$$= \omega_1\left(f, \frac{1}{n^\alpha} \right) + 2e^4 \|f\|_\infty\, e^{-2n^{(1-\alpha)}}.$$

That is

$$\left| \sum_{k=\lceil na \rceil}^{\lfloor nb \rfloor} \left(f\left(\frac{k}{n}\right) - f(x) \right) \Psi(nx - k) \right| \leq \omega_1\left(f, \frac{1}{n^\alpha}\right) + 2e^4 \|f\|_\infty \, e^{-2n^{(1-\alpha)}}.$$

Using (2.6) we prove the claim. ∎

Next we give

Theorem 2.15. *Let $f \in C_B(\mathbb{R})$, $0 < \alpha < 1$, $n \in \mathbb{N}$, $x \in \mathbb{R}$. Then*
i)

$$\left| \overline{F}_n(f, x) - f(x) \right| \leq \omega_1\left(f, \frac{1}{n^\alpha}\right) + 2e^4 \|f\|_\infty \, e^{-2n^{(1-\alpha)}} =: \mu, \qquad (2.25)$$

and
ii)

$$\left\| \overline{F}_n(f) - f \right\|_\infty \leq \mu. \qquad (2.26)$$

Proof. We see that

$$\left| \overline{F}_n(f, x) - f(x) \right| = \left| \sum_{k=-\infty}^{\infty} f\left(\frac{k}{n}\right) \Psi(nx - k) - f(x) \sum_{k=-\infty}^{\infty} \Psi(nx - k) \right| =$$

$$\left| \sum_{k=-\infty}^{\infty} \left(f\left(\frac{k}{n}\right) - f(x) \right) \Psi(nx - k) \right| \leq \sum_{k=-\infty}^{\infty} \left| f\left(\frac{k}{n}\right) - f(x) \right| \Psi(nx - k) =$$

$$\sum_{\substack{k=-\infty \\ \left| \frac{k}{n} - x \right| \leq \frac{1}{n^\alpha}}}^{\infty} \left| f\left(\frac{k}{n}\right) - f(x) \right| \Psi(nx - k) +$$

$$\sum_{\substack{k=-\infty \\ \left| \frac{k}{n} - x \right| > \frac{1}{n^\alpha}}}^{\infty} \left| f\left(\frac{k}{n}\right) - f(x) \right| \Psi(nx - k) \leq$$

$$\sum_{\substack{k=-\infty \\ \left| \frac{k}{n} - x \right| \leq \frac{1}{n^\alpha}}}^{\infty} \omega_1\left(f, \left| \frac{k}{n} - x \right|\right) \Psi(nx - k) +$$

$$2\|f\|_\infty \sum_{\substack{k=-\infty \\ \left| \frac{k}{n} - x \right| > \frac{1}{n^\alpha}}}^{\infty} \Psi(nx - k) \leq$$

$$\omega_1\left(f, \frac{1}{n^\alpha}\right) \sum_{\substack{k=-\infty \\ \left|\frac{k}{n} - x\right| \le \frac{1}{n^\alpha}}}^{\infty} \Psi(nx - k) +$$

$$2\|f\|_\infty \sum_{\substack{k=-\infty \\ |k - nx| > n^{1-\alpha}}}^{\infty} \Psi(nx - k) \le$$

$$\omega_1\left(f, \frac{1}{n^\alpha}\right) + 2\|f\|_\infty e^4 e^{-2n^{(1-\alpha)}}$$

$$= \omega_1\left(f, \frac{1}{n^\alpha}\right) + 2e^4 \|f\|_\infty e^{-2n^{(1-\alpha)}},$$

proving the claim. ∎

In the next we discuss high order of approximation by using the smoothness of f.

Theorem 2.16. Let $f \in C^N([a,b])$, $n, N \in \mathbb{N}$, $0 < \alpha < 1$, $x \in [a,b]$. Then
i)

$$|F_n(f, x) - f(x)| \le (4.1488766) \cdot \tag{2.27}$$

$$\left\{ \sum_{j=1}^{N} \frac{|f^{(j)}(x)|}{j!} \left[\frac{1}{n^{\alpha j}} + e^4 (b-a)^j e^{-2n^{(1-\alpha)}} \right] + \right.$$

$$\left. \left[\omega_1\left(f^{(N)}, \frac{1}{n^\alpha}\right) \frac{1}{n^{\alpha N} N!} + \frac{2e^4 \|f^{(N)}\|_\infty (b-a)^N}{N!} e^{-2n^{(1-\alpha)}} \right] \right\},$$

ii) suppose further $f^{(j)}(x_0) = 0$, $j = 1, ..., N$, for some $x_0 \in [a, b]$, it holds

$$|F_n(f, x_0) - f(x_0)| \le (4.1488766) \cdot \tag{2.28}$$

$$\left[\omega_1\left(f^{(N)}, \frac{1}{n^\alpha}\right) \frac{1}{n^{\alpha N} N!} + \frac{2e^4 \|f^{(N)}\|_\infty (b-a)^N}{N!} e^{-2n^{(1-\alpha)}} \right],$$

notice here the extremely high rate of convergence at $n^{-(N+1)\alpha}$,
iii)

$$\|F_n(f) - f\|_\infty \le (4.1488766) \cdot \tag{2.29}$$

$$\left\{ \sum_{j=1}^{N} \frac{\|f^{(j)}\|_\infty}{j!} \left[\frac{1}{n^{\alpha j}} + e^4 (b-a)^j e^{-2n^{(1-\alpha)}} \right] + \right.$$

$$\left. \left[\omega_1\left(f^{(N)}, \frac{1}{n^\alpha}\right) \frac{1}{n^{\alpha N} N!} + \frac{2e^4 \|f^{(N)}\|_\infty (b-a)^N}{N!} e^{-2n^{(1-\alpha)}} \right] \right\}.$$

Proof. Next we apply Taylor's formula with integral remainder.

We have (here $\frac{k}{n}, x \in [a, b]$)

$$f\left(\frac{k}{n}\right) = \sum_{j=0}^{N} \frac{f^{(j)}(x)}{j!} \left(\frac{k}{n} - x\right)^j + \int_x^{\frac{k}{n}} \left(f^{(N)}(t) - f^{(N)}(x)\right) \frac{\left(\frac{k}{n} - t\right)^{N-1}}{(N-1)!} dt.$$

Then

$$f\left(\frac{k}{n}\right) \Psi(nx - k) = \sum_{j=0}^{N} \frac{f^{(j)}(x)}{j!} \Psi(nx - k) \left(\frac{k}{n} - x\right)^j +$$

$$\Psi(nx - k) \int_x^{\frac{k}{n}} \left(f^{(N)}(t) - f^{(N)}(x)\right) \frac{\left(\frac{k}{n} - t\right)^{N-1}}{(N-1)!} dt.$$

Hence

$$\sum_{k=\lceil na \rceil}^{\lfloor nb \rfloor} f\left(\frac{k}{n}\right) \Psi(nx - k) - f(x) \sum_{k=\lceil na \rceil}^{\lfloor nb \rfloor} \Psi(nx - k) =$$

$$\sum_{j=1}^{N} \frac{f^{(j)}(x)}{j!} \sum_{k=\lceil na \rceil}^{\lfloor nb \rfloor} \Psi(nx - k) \left(\frac{k}{n} - x\right)^j +$$

$$\sum_{k=\lceil na \rceil}^{\lfloor nb \rfloor} \Psi(nx - k) \int_x^{\frac{k}{n}} \left(f^{(N)}(t) - f^{(N)}(x)\right) \frac{\left(\frac{k}{n} - t\right)^{N-1}}{(N-1)!} dt.$$

Therefore

$$F_n^*(f, x) - f(x) \left(\sum_{k=\lceil na \rceil}^{\lfloor nb \rfloor} \Psi(nx - k)\right) = \sum_{j=1}^{N} \frac{f^{(j)}(x)}{j!} F_n^*\left((\cdot - x)^j\right) + \Lambda_n(x),$$

where

$$\Lambda_n(x) := \sum_{k=\lceil na \rceil}^{\lfloor nb \rfloor} \Psi(nx - k) \int_x^{\frac{k}{n}} \left(f^{(N)}(t) - f^{(N)}(x)\right) \frac{\left(\frac{k}{n} - t\right)^{N-1}}{(N-1)!} dt.$$

We suppose that $b - a > \frac{1}{n^\alpha}$, which is always the case for large enough $n \in \mathbb{N}$, that is when $n > \left\lceil (b-a)^{-\frac{1}{\alpha}} \right\rceil$.

Thus $\left|\frac{k}{n} - x\right| \le \frac{1}{n^\alpha}$ or $\left|\frac{k}{n} - x\right| > \frac{1}{n^\alpha}$.

As in [2], pp. 72-73 for

$$\gamma := \int_x^{\frac{k}{n}} \left(f^{(N)}(t) - f^{(N)}(x)\right) \frac{\left(\frac{k}{n} - t\right)^{N-1}}{(N-1)!} dt,$$

in case of $\left| \frac{k}{n} - x \right| \le \frac{1}{n^\alpha}$, we find that

$$|\gamma| \le \omega_1 \left(f^{(N)}, \frac{1}{n^\alpha} \right) \frac{1}{n^{\alpha N} N!}$$

(for $x \le \frac{k}{n}$ or $x \ge \frac{k}{n}$).

Notice also for $x \le \frac{k}{n}$ that

$$\left| \int_x^{\frac{k}{n}} \left(f^{(N)}(t) - f^{(N)}(x) \right) \frac{\left(\frac{k}{n} - t \right)^{N-1}}{(N-1)!} dt \right| \le$$

$$\int_x^{\frac{k}{n}} \left| f^{(N)}(t) - f^{(N)}(x) \right| \frac{\left(\frac{k}{n} - t \right)^{N-1}}{(N-1)!} dt \le$$

$$2 \left\| f^{(N)} \right\|_\infty \int_x^{\frac{k}{n}} \frac{\left(\frac{k}{n} - t \right)^{N-1}}{(N-1)!} dt = 2 \left\| f^{(N)} \right\|_\infty \frac{\left(\frac{k}{n} - x \right)^N}{N!} \le 2 \left\| f^{(N)} \right\|_\infty \frac{(b-a)^N}{N!}.$$

Next suppose $\frac{k}{n} \le x$, then

$$\left| \int_x^{\frac{k}{n}} \left(f^{(N)}(t) - f^{(N)}(x) \right) \frac{\left(\frac{k}{n} - t \right)^{N-1}}{(N-1)!} dt \right| =$$

$$\left| \int_{\frac{k}{n}}^x \left(f^{(N)}(t) - f^{(N)}(x) \right) \frac{\left(\frac{k}{n} - t \right)^{N-1}}{(N-1)!} dt \right| \le$$

$$\int_{\frac{k}{n}}^x \left| f^{(N)}(t) - f^{(N)}(x) \right| \frac{\left(t - \frac{k}{n} \right)^{N-1}}{(N-1)!} dt \le$$

$$2 \left\| f^{(N)} \right\|_\infty \int_{\frac{k}{n}}^x \frac{\left(t - \frac{k}{n} \right)^{N-1}}{(N-1)!} dt = 2 \left\| f^{(N)} \right\|_\infty \frac{\left(x - \frac{k}{n} \right)^N}{N!} \le 2 \left\| f^{(N)} \right\|_\infty \frac{(b-a)^N}{N!}.$$

Thus

$$|\gamma| \le 2 \left\| f^{(N)} \right\|_\infty \frac{(b-a)^N}{N!},$$

in all two cases.

Therefore

$$\Lambda_n(x) = \sum_{\substack{k=\lceil na \rceil \\ \left| \frac{k}{n} - x \right| \le \frac{1}{n^\alpha}}}^{\lfloor nb \rfloor} \Psi(nx - k) \gamma + \sum_{\substack{k=\lceil na \rceil \\ \left| \frac{k}{n} - x \right| > \frac{1}{n^\alpha}}}^{\lfloor nb \rfloor} \Psi(nx - k) \gamma.$$

Hence

$$|\Lambda_n(x)| \le \sum_{\substack{k=\lceil na \rceil \\ \left| \frac{k}{n} - x \right| \le \frac{1}{n^\alpha}}}^{\lfloor nb \rfloor} \Psi(nx - k) \left(\omega_1 \left(f^{(N)}, \frac{1}{n^\alpha} \right) \frac{1}{N! n^{N\alpha}} \right) +$$

$$\left(\sum_{\substack{k=\lceil na \rceil \\ |\frac{k}{n}-x|>\frac{1}{n^{\alpha}}}}^{\lfloor nb \rfloor} \Psi\left(nx-k\right) \right) 2 \left\| f^{(N)} \right\|_{\infty} \frac{(b-a)^N}{N!} \le$$

$$\omega_1\left(f^{(N)}, \frac{1}{n^{\alpha}}\right) \frac{1}{N! n^{N\alpha}} + 2\left\| f^{(N)} \right\|_{\infty} \frac{(b-a)^N}{N!} e^4 e^{-2n^{(1-\alpha)}}.$$

Consequently we have

$$|A_n\left(x\right)| \le \omega_1\left(f^{(N)}, \frac{1}{n^{\alpha}}\right) \frac{1}{n^{\alpha N} N!} + 2e^4 \frac{\left\| f^{(N)} \right\|_{\infty} (b-a)^N}{N!} e^{-2n^{(1-\alpha)}}.$$

We further observe that

$$F_n^*\left((\cdot - x)^j\right) = \sum_{k=\lceil na \rceil}^{\lfloor nb \rfloor} \Psi\left(nx-k\right) \left(\frac{k}{n}-x\right)^j.$$

Therefore

$$\left| F_n^*\left((\cdot - x)^j\right) \right| \le \sum_{k=\lceil na \rceil}^{\lfloor nb \rfloor} \Psi\left(nx-k\right) \left| \frac{k}{n}-x \right|^j =$$

$$\sum_{\substack{k=\lceil na \rceil \\ |\frac{k}{n}-x| \le \frac{1}{n^{\alpha}}}}^{\lfloor nb \rfloor} \Psi\left(nx-k\right) \left| \frac{k}{n}-x \right|^j + \sum_{\substack{k=\lceil na \rceil \\ |\frac{k}{n}-x| > \frac{1}{n^{\alpha}}}}^{\lfloor nb \rfloor} \Psi\left(nx-k\right) \left| \frac{k}{n}-x \right|^j \le$$

$$\frac{1}{n^{\alpha j}} \sum_{\substack{k=\lceil na \rceil \\ |\frac{k}{n}-x| \le \frac{1}{n^{\alpha}}}}^{\lfloor nb \rfloor} \Psi\left(nx-k\right) + (b-a)^j \sum_{\substack{k=\lceil na \rceil \\ |k-nx| > n^{1-\alpha}}}^{\lfloor nb \rfloor} \Psi\left(nx-k\right)$$

$$\le \frac{1}{n^{\alpha j}} + (b-a)^j e^4 e^{-2n^{(1-\alpha)}}.$$

Hence

$$\left| F_n^*\left((\cdot - x)^j\right) \right| \le \frac{1}{n^{\alpha j}} + (b-a)^j e^4 e^{-2n^{(1-\alpha)}},$$

for $j = 1, ..., N$.

Putting things together we have proved

$$|F_n^*\left(f, x\right) - f\left(x\right)| \le \sum_{j=1}^{N} \frac{\left| f^{(j)}\left(x\right) \right|}{j!} \left[\frac{1}{n^{\alpha j}} + e^4 (b-a)^j e^{-2n^{(1-\alpha)}} \right] +$$

$$\left[\omega_1 \left(f^{(N)}, \frac{1}{n^\alpha} \right) \frac{1}{n^{\alpha N} N!} + \frac{2e^4 \left\| f^{(N)} \right\|_\infty (b-a)^N}{N!} e^{-2n^{(1-\alpha)}} \right],$$

that is establishing theorem.   ■

We make

Remark 2.17. *We notice that*

$$F_n(f,x) - \sum_{j=1}^N \frac{f^{(j)}(x)}{j!} F_n\left((\cdot - x)^j \right) - f(x) =$$

$$\frac{F_n^*(f,x)}{\left(\sum_{k=\lceil na \rceil}^{\lfloor nb \rfloor} \Psi(nx-k) \right)} - \frac{1}{\left(\sum_{k=\lceil na \rceil}^{\lfloor nb \rfloor} \Psi(nx-k) \right)} \left(\sum_{j=1}^n \frac{f^{(j)}(x)}{j!} F_n^*\left((\cdot - x)^j \right) \right)$$

$$-f(x) = \frac{1}{\left(\sum_{k=\lceil na \rceil}^{\lfloor nb \rfloor} \Psi(nx-k) \right)} \cdot$$

$$\left[F_n^*(f,x) - \left(\sum_{j=1}^n \frac{f^{(j)}(x)}{j!} F_n^*\left((\cdot - x)^j \right) \right) - \left(\sum_{k=\lceil na \rceil}^{\lfloor nb \rfloor} \Psi(nx-k) \right) f(x) \right].$$

Therefore we find

$$\left| F_n(f,x) - \sum_{j=1}^N \frac{f^{(j)}(x)}{j!} F_n\left((\cdot - x)^j \right) - f(x) \right| \le (4.1488766) \cdot$$

$$\left| F_n^*(f,x) - \left(\sum_{j=1}^n \frac{f^{(j)}(x)}{j!} F_n^*\left((\cdot - x)^j \right) \right) - \left(\sum_{k=\lceil na \rceil}^{\lfloor nb \rfloor} \Psi(nx-k) \right) f(x) \right|,$$

$$(2.30)$$

$\forall\, x \in [a,b].$

In the next three Theorems 2.18-2.20 we present more general and flexible upper bounds to our error quantities.

We give

Theorem 2.18. *Let* $f \in C^N([a,b])$, $n, N \in \mathbb{N}$, $0 < \alpha < 1$, $x \in [a,b]$. *Then*
1)

$$\left| F_n(f,x) - \sum_{j=1}^N \frac{\left(f \circ \ln_{\frac{1}{e}} \right)^{(j)} (e^{-x})}{j!} F_n\left((e^{-\cdot} - e^{-x})^j, x \right) - f(x) \right| \le$$

$$(4.1488766) \left[\frac{e^{-\alpha N}}{N! n^{N\alpha}} \omega_1 \left(\left(f \circ \ln_{\frac{1}{e}} \right)^{(N)}, \frac{e^{-\alpha}}{n^\alpha} \right) + \right.$$

$$\frac{2e^4 \left(e^{-a} - e^{-b}\right)^N}{N!} \left\| \left(f \circ \ln_{\frac{1}{e}}\right)^{(N)} \right\|_{\infty} e^{-2n^{(1-\alpha)}} \right], \qquad (2.31)$$

2)
$$|F_n(f, x) - f(x)| \le (4.1488766) \cdot \qquad (2.32)$$

$$\left\{ \sum_{j=1}^{N} \frac{\left| \left(f \circ \ln_{\frac{1}{e}}\right)^{(j)} (e^{-x}) \right|}{j!} e^{-aj} \left[\frac{1}{n^{\alpha j}} + e^4 (b-a)^j e^{-2n^{(1-\alpha)}} \right] + \right.$$

$$\left[\frac{e^{-aN}}{N! n^{\alpha N}} \omega_1 \left(\left(f \circ \ln_{\frac{1}{e}}\right)^{(N)}, \frac{e^{-a}}{n^\alpha} \right) + \right.$$

$$\left. \frac{2e^4 \left(e^{-a} - e^{-b}\right)^N}{N!} \left\| \left(f \circ \ln_{\frac{1}{e}}\right)^{(N)} \right\|_{\infty} e^{-2n^{(1-\alpha)}} \right] \right\},$$

3) If $f^{(j)}(x_0) = 0$, $j = 1, ..., N$, it holds
$$|F_n(f, x_0) - f(x_0)| \le (4.1488766) \cdot \qquad (2.33)$$

$$\left[\frac{e^{-aN}}{N! n^{N\alpha}} \omega_1 \left(\left(f \circ \ln_{\frac{1}{e}}\right)^{(N)}, \frac{e^{-a}}{n^\alpha} \right) + \right.$$

$$\left. \frac{2e^4 \left(e^{-a} - e^{-b}\right)^N}{N!} \left\| \left(f \circ \ln_{\frac{1}{e}}\right)^{(N)} \right\|_{\infty} e^{-2n^{(1-\alpha)}} \right].$$

Observe here the speed of convergence is extremely high at $\frac{1}{n^{(N+1)\alpha}}$.

Proof. Call $F := f \circ \ln_{\frac{1}{e}}$. Let $x, \frac{k}{n} \in [a, b]$. Then

$$f\left(\frac{k}{n}\right) - f(x) = \sum_{j=1}^{N} \frac{F^{(j)} (e^{-x})}{j!} \left(e^{-\frac{k}{n}} - e^{-x}\right)^j + \overline{K}_N \left(x, \frac{k}{n}\right),$$

where
$$\overline{K}_N \left(x, \frac{k}{n}\right) := -\frac{1}{(N-1)!}.$$

$$\int_x^{\frac{k}{n}} \left(e^{-\frac{k}{n}} - e^{-s}\right)^{N-1} \left[F^{(N)} (e^{-s}) - F^{(N)} (e^{-x})\right] e^{-s} ds.$$

Thus
$$\sum_{k=\lceil na \rceil}^{\lfloor nb \rfloor} \Psi(nx - k) f\left(\frac{k}{n}\right) - f(x) \sum_{k=\lceil na \rceil}^{\lfloor nb \rfloor} \Psi(nx - k) =$$

$$\sum_{j=1}^{N} \frac{F^{(j)} (e^{-x})}{j!} \sum_{k=\lceil na \rceil}^{\lfloor nb \rfloor} \Psi(nx - k) \left(e^{-\frac{k}{n}} - e^{-x}\right)^j + \sum_{k=\lceil na \rceil}^{\lfloor nb \rfloor} \Psi(nx - k) \frac{1}{(N-1)!}.$$

$$\int_x^{\frac{k}{n}} \left(e^{-\frac{k}{n}} - e^{-s} \right)^{N-1} \left[F^{(N)} \left(e^{-s} \right) - F^{(N)} \left(e^{-x} \right) \right] de^{-s}.$$

Therefore

$$F_n^* \left(f, x \right) - f \left(x \right) \left(\sum_{k=\lceil na \rceil}^{\lfloor nb \rfloor} \Psi \left(nx - k \right) \right) =$$

$$\sum_{j=1}^N \frac{F^{(j)} \left(e^{-x} \right)}{j!} F_n^* \left(\left(e^{-\cdot} - e^{-x} \right)^j, x \right) + U_n \left(x \right),$$

where

$$U_n \left(x \right) := \sum_{k=\lceil na \rceil}^{\lfloor nb \rfloor} \Psi \left(nx - k \right) \mu,$$

with

$$\mu := \frac{1}{(N-1)!} \int_{e^{-x}}^{e^{-\frac{k}{n}}} \left(e^{-\frac{k}{n}} - w \right)^{N-1} \left[F^{(N)} \left(w \right) - F^{(N)} \left(e^{-x} \right) \right] dw.$$

Case of $\left| \frac{k}{n} - x \right| \le \frac{1}{n^\alpha}$.
i) Subcase of $x \ge \frac{k}{n}$. I.e. $e^{-\frac{k}{n}} \ge e^{-x}$.

$$|\mu| \le \frac{1}{(N-1)!} \int_{e^{-x}}^{e^{-\frac{k}{n}}} \left(e^{-\frac{k}{n}} - w \right)^{N-1} \left| F^{(N)} \left(w \right) - F^{(N)} \left(e^{-x} \right) \right| dw \le$$

$$\frac{1}{(N-1)!} \int_{e^{-x}}^{e^{-\frac{k}{n}}} \left(e^{-\frac{k}{n}} - w \right)^{N-1} \omega_1 \left(F^{(N)}, \left| w - e^{-x} \right| \right) dw \le$$

$$\frac{1}{(N-1)!} \omega_1 \left(F^{(N)}, \left| e^{-\frac{k}{n}} - e^{-x} \right| \right) \int_{e^{-x}}^{e^{-\frac{k}{n}}} \left(e^{-\frac{k}{n}} - w \right)^{N-1} dw \le$$

$$\frac{\left(e^{-\frac{k}{n}} - e^{-x} \right)^N}{N!} \omega_1 \left(F^{(N)}, e^{-a} \left| x - \frac{k}{n} \right| \right) \le$$

$$e^{-aN} \frac{\left| x - \frac{k}{n} \right|^N}{N!} \omega_1 \left(F^{(N)}, e^{-a} \left| x - \frac{k}{n} \right| \right) \le$$

$$\frac{e^{-aN}}{N! n^{\alpha N}} \omega_1 \left(F^{(N)}, \frac{e^{-a}}{n^\alpha} \right).$$

Hence when $x \ge \frac{k}{n}$ we derive

$$|\mu| \le \frac{e^{-aN}}{N! n^{\alpha N}} \omega_1 \left(F^{(N)}, \frac{e^{-a}}{n^\alpha} \right).$$

ii) Subcase of $\frac{k}{n} \geq x$. Then $e^{-\frac{k}{n}} \leq e^{-x}$ and

$$|\mu| \leq \frac{1}{(N-1)!} \int_{e^{-\frac{k}{n}}}^{e^{-x}} \left(w - e^{-\frac{k}{n}} \right)^{N-1} \left| F^{(N)}(w) - F^{(N)}\left(e^{-x}\right) \right| dw \leq$$

$$\frac{1}{(N-1)!} \int_{e^{-\frac{k}{n}}}^{e^{-x}} \left(w - e^{-\frac{k}{n}} \right)^{N-1} \omega_1 \left(F^{(N)}, \left| w - e^{-x} \right| \right) dw \leq$$

$$\frac{1}{(N-1)!} \omega_1 \left(F^{(N)}, e^{-x} - e^{-\frac{k}{n}} \right) \int_{e^{-\frac{k}{n}}}^{e^{-x}} \left(w - e^{-\frac{k}{n}} \right)^{N-1} dw \leq$$

$$\frac{1}{(N-1)!} \omega_1 \left(F^{(N)}, e^{-a} \left| x - \frac{k}{n} \right| \right) \frac{\left(e^{-x} - e^{-\frac{k}{n}} \right)^N}{N} \leq$$

$$\frac{1}{N!} \omega_1 \left(F^{(N)}, \frac{e^{-a}}{n^\alpha} \right) e^{-aN} \left| x - \frac{k}{n} \right|^N \leq$$

$$\frac{1}{N!} \omega_1 \left(F^{(N)}, \frac{e^{-a}}{n^\alpha} \right) \frac{e^{-aN}}{n^{\alpha N}},$$

i.e.

$$|\mu| \leq \frac{e^{-aN}}{N! n^{\alpha N}} \omega_1 \left(F^{(N)}, \frac{e^{-a}}{n^\alpha} \right),$$

when $\frac{k}{n} \geq x$. So in general when $\left| \frac{k}{n} - x \right| \leq \frac{1}{n^\alpha}$ we proved that

$$|\mu| \leq \frac{e^{-aN}}{N! n^{\alpha N}} \omega_1 \left(F^{(N)}, \frac{e^{-a}}{n^\alpha} \right).$$

Also we observe:
i)' When $\frac{k}{n} \leq x$, we obtain

$$|\mu| \leq \frac{1}{(N-1)!} \left(\int_{e^{-x}}^{e^{-\frac{k}{n}}} \left(e^{-\frac{k}{n}} - w \right)^{N-1} dw \right) 2 \left\| F^{(N)} \right\|_\infty$$

$$= \frac{2 \left\| F^{(N)} \right\|_\infty}{(N-1)!} \frac{\left(e^{-\frac{k}{n}} - e^{-x} \right)^N}{N} \leq \frac{2 \left\| F^{(N)} \right\|_\infty}{N!} \left(e^{-a} - e^{-b} \right)^N.$$

ii)' When $\frac{k}{n} \geq x$, we get

$$|\mu| \leq \frac{1}{(N-1)!} \left(\int_{e^{-\frac{k}{n}}}^{e^{-x}} \left(w - e^{-\frac{k}{n}} \right)^{N-1} dw \right) 2 \left\| F^{(N)} \right\|_\infty$$

$$= \frac{2 \left\| F^{(N)} \right\|_\infty}{N!} \left(e^{-x} - e^{-\frac{k}{n}} \right)^N \leq \frac{2 \left\| F^{(N)} \right\|_\infty}{N!} \left(e^{-a} - e^{-b} \right)^N.$$

We proved always true that

$$|\mu| \leq \frac{2\left(e^{-a} - e^{-b}\right)^N \left\|F^{(N)}\right\|_\infty}{N!}.$$

Consequently we derive

$$|U_n(x)| \leq \sum_{\substack{k = \lceil na \rceil \\ \left|\frac{k}{n} - x\right| \leq \frac{1}{n^\alpha}}}^{\lfloor nb \rfloor} \Psi(nx - k)|\mu| + \sum_{\substack{k = \lceil na \rceil \\ \left|\frac{k}{n} - x\right| > \frac{1}{n^\alpha}}}^{\lfloor nb \rfloor} \Psi(nx - k)|\mu| \leq$$

$$\left(\frac{e^{-aN}}{N! n^{\alpha N}}\omega_1\left(F^{(N)}, \frac{e^{-a}}{n^\alpha}\right)\right)\left(\sum_{k=-\infty}^{\infty} \Psi(nx - k)\right) +$$

$$\left(\frac{2\left(e^{-a} - e^{-b}\right)^N \left\|F^{(N)}\right\|_\infty}{N!}\right)\left(\sum_{\substack{k = -\infty \\ |k - nx| > n^{1-\alpha}}}^{\infty} \Psi(nx - k)\right) \leq$$

$$\frac{e^{-aN}}{N! n^{\alpha N}}\omega_1\left(F^{(N)}, \frac{e^{-a}}{n^\alpha}\right) + \left(\frac{2\left(e^{-a} - e^{-b}\right)^N \left\|F^{(N)}\right\|_\infty}{N!}\right)e^4 e^{-2n^{(1-\alpha)}}.$$

So we have proved that

$$|U_n(x)| \leq \frac{e^{-aN}}{N! n^{\alpha N}}\omega_1\left(F^{(N)}, \frac{e^{-a}}{n^\alpha}\right) + \frac{2e^4\left(e^{-a} - e^{-b}\right)^N \left\|F^{(N)}\right\|_\infty}{N!}e^{-2n^{(1-\alpha)}}.$$

We also notice that

$$\left|F_n^*\left(\left(e^{-\cdot} - e^{-x}\right)^j, x\right)\right| \leq F_n^*\left(\left|e^{-\cdot} - e^{-x}\right|^j, x\right) \leq$$

$$e^{-aj}F_n^*\left(|\cdot - x|^j, x\right) = e^{-aj}\left(\sum_{k=\lceil na \rceil}^{\lfloor nb \rfloor} \Psi(nx - k)\left|\frac{k}{n} - x\right|^j\right) = e^{-aj}.$$

$$\left[\sum_{\substack{k = \lceil na \rceil \\ \left|x - \frac{k}{n}\right| \leq \frac{1}{n^\alpha}}}^{\lfloor nb \rfloor} \Psi(nx - k)\left|\frac{k}{n} - x\right|^j + \sum_{\substack{k = \lceil na \rceil \\ \left|x - \frac{k}{n}\right| > \frac{1}{n^\alpha}}}^{\lfloor nb \rfloor} \Psi(nx - k)\left|\frac{k}{n} - x\right|^j\right] \leq$$

$$e^{-aj} \left[\frac{1}{n^{\alpha j}} + (b-a)^j \sum_{\substack{k = \lceil na \rceil \\ \left| x - \frac{k}{n} \right| > \frac{1}{n^\alpha}}}^{\lfloor nb \rfloor} \Psi \left(nx - k \right) \right] \le$$

$$e^{-aj} \left[\frac{1}{n^{\alpha j}} + e^4 (b-a)^j e^{-2n^{(1-\alpha)}} \right].$$

Thus we have established

$$\left| F_n^* \left(\left(e^{-\cdot} - e^{-x} \right)^j, x \right) \right| \le e^{-aj} \left[\frac{1}{n^{\alpha j}} + e^4 (b-a)^j e^{-2n^{(1-\alpha)}} \right],$$

for $j = 1, ..., N$, and the theorem.   ∎

We continue with

Theorem 2.19. *Let* $f \in C^N \left(\left[-\frac{\pi}{2} + \varepsilon, \frac{\pi}{2} - \varepsilon \right] \right)$, $n, N \in \mathbb{N}$, $0 < \varepsilon < \frac{\pi}{2}$, ε *small,* $x \in \left[-\frac{\pi}{2} + \varepsilon, \frac{\pi}{2} - \varepsilon \right]$, $0 < \alpha < 1$. *Then*
   1)

$$\left| F_n (f, x) - \sum_{j=1}^N \frac{\left(f \circ \sin^{-1} \right)^{(j)} (\sin x)}{j!} F_n \left(\left(\sin \cdot - \sin x \right)^j, x \right) - f(x) \right| \le$$

(2.34)

$$(4.1488766) \left[\frac{\omega_1 \left(\left(f \circ \sin^{-1} \right)^{(N)}, \frac{1}{n^\alpha} \right)}{n^{\alpha N} N!} + \right.$$

$$\left. \left(\frac{e^4 2^{N+1} \left\| \left(f \circ \sin^{-1} \right)^{(N)} \right\|_\infty}{N!} \right) e^{-2n^{(1-\alpha)}} \right],$$

   2)

$$|F_n (f, x) - f(x)| \le (4.1488766) \cdot$$

(2.35)

$$\left\{ \sum_{j=1}^N \frac{\left| \left(f \circ \sin^{-1} \right)^{(j)} (\sin x) \right|}{j!} \left[\frac{1}{n^{\alpha j}} + e^4 (\pi - 2\varepsilon)^j e^{-2n^{(1-\alpha)}} \right] + \right.$$

$$\left. \left[\frac{\omega_1 \left(\left(f \circ \sin^{-1} \right)^{(N)}, \frac{1}{n^\alpha} \right)}{N! n^{\alpha N}} + \left(\frac{e^4 2^{N+1}}{N!} \left\| \left(f \circ \sin^{-1} \right)^{(N)} \right\|_\infty \right) e^{-2n^{(1-\alpha)}} \right] \right\},$$

   3) suppose further $f^{(j)} (x_0) = 0$, $j = 1, ..., N$ *for some* $x_0 \in \left[-\frac{\pi}{2} + \varepsilon, \frac{\pi}{2} - \varepsilon \right]$, *it holds*

$$|F_n (f, x_0) - f(x_0)| \le (4.1488766) \cdot$$

(2.36)

$$\left[\frac{\omega_1\left(\left(f \circ \sin^{-1}\right)^{(N)}, \frac{1}{n^\alpha}\right)}{n^{\alpha N} N!} + \left(\frac{e^4 2^{N+1} \left\| \left(f \circ \sin^{-1}\right)^{(N)} \right\|_\infty}{N!} \right) e^{-2n^{(1-\alpha)}} \right].$$

Notice in the last the high speed of convergence of order $n^{-\alpha(N+1)}$.

Proof. Call $F := f \circ \sin^{-1}$ and let $\frac{k}{n}, x \in \left[-\frac{\pi}{2} + \varepsilon, \frac{\pi}{2} - \varepsilon\right]$. Then

$$f\left(\frac{k}{n}\right) - f(x) = \sum_{j=1}^{N} \frac{F^{(j)}(\sin x)}{j!} \left(\sin\frac{k}{n} - \sin x\right)^j +$$

$$\frac{1}{(N-1)!} \int_x^{\frac{k}{n}} \left(\sin\frac{k}{n} - \sin s\right)^{N-1} \left(F^{(N)}(\sin s) - F^{(N)}(\sin x)\right) d\sin s.$$

Hence

$$\sum_{k=\lceil na\rceil}^{\lfloor nb\rfloor} f\left(\frac{k}{n}\right) \Psi(nx-k) - f(x) \sum_{k=\lceil na\rceil}^{\lfloor nb\rfloor} \Psi(nx-k) =$$

$$\sum_{j=1}^{N} \frac{F^{(j)}(\sin x)}{j!} \sum_{k=\lceil na\rceil}^{\lfloor nb\rfloor} \Psi(nx-k) \left(\sin\frac{k}{n} - \sin x\right)^j + \frac{1}{(N-1)!} \sum_{k=\lceil na\rceil}^{\lfloor nb\rfloor} \Psi(nx-k) \cdot$$

$$\int_x^{\frac{k}{n}} \left(\sin\frac{k}{n} - \sin s\right)^{N-1} \left(F^{(N)}(\sin s) - F^{(N)}(\sin x)\right) d\sin s.$$

Set here $a = -\frac{\pi}{2} + \varepsilon$, $b = \frac{\pi}{2} - \varepsilon$. Thus

$$F_n^*(f, x) - f(x) \sum_{k=\lceil na\rceil}^{\lfloor nb\rfloor} \Psi(nx-k) =$$

$$\sum_{j=1}^{N} \frac{F^{(j)}(\sin x)}{j!} F_n^* \left((\sin \cdot - \sin x)^j, x\right) + M_n(x),$$

where

$$M_n(x) := \sum_{k=\lceil na\rceil}^{\lfloor nb\rfloor} \Psi(nx-k) \rho,$$

with

$$\rho := \frac{1}{(N-1)!} \int_x^{\frac{k}{n}} \left(\sin\frac{k}{n} - \sin s\right)^{N-1} \left(F^{(N)}(\sin s) - F^{(N)}(\sin x)\right) d\sin s.$$

Case of $\left|\frac{k}{n} - x\right| \leq \frac{1}{n^\alpha}$.

i) Subcase of $\frac{k}{n} \geq x$. The function sin is increasing on $[a, b]$, i.e. $\sin\frac{k}{n} \geq \sin x$.

Then

$$|\rho| \leq \frac{1}{(N-1)!} \int_x^{\frac{k}{n}} \left(\sin \frac{k}{n} - \sin s\right)^{N-1} \left|F^{(N)}(\sin s) - F^{(N)}(\sin x)\right| d\sin s =$$

$$\frac{1}{(N-1)!} \int_{\sin x}^{\sin \frac{k}{n}} \left(\sin \frac{k}{n} - w\right)^{N-1} \left(F^{(N)}(w) - F^{(N)}(\sin x)\right) dw \leq$$

$$\frac{1}{(N-1)!} \int_{\sin x}^{\sin \frac{k}{n}} \left(\sin \frac{k}{n} - w\right)^{N-1} \omega_1 \left(F^{(N)}, |w - \sin x|\right) dw \leq$$

$$\omega_1 \left(F^{(N)}, \left|\sin \frac{k}{n} - \sin x\right|\right) \frac{\left(\sin \frac{k}{n} - \sin x\right)^N}{N!} \leq$$

$$\omega_1 \left(F^{(N)}, \left|\frac{k}{n} - x\right|\right) \frac{\left(\frac{k}{n} - x\right)^N}{N!} \leq \omega_1 \left(F^{(N)}, \frac{1}{n^\alpha}\right) \frac{1}{n^{\alpha N} N!}.$$

So if $\frac{k}{n} \geq x$, then

$$|\rho| \leq \omega_1 \left(F^{(N)}, \frac{1}{n^\alpha}\right) \frac{1}{N! n^{\alpha N}}.$$

ii) Subcase of $\frac{k}{n} \leq x$, then $\sin \frac{k}{n} \leq \sin x$. Hence

$$|\rho| \leq \frac{1}{(N-1)!} \int_{\frac{k}{n}}^x \left(\sin s - \sin \frac{k}{n}\right)^{N-1} \left|F^{(N)}(\sin s) - F^{(N)}(\sin x)\right| d\sin s =$$

$$\frac{1}{(N-1)!} \int_{\sin \frac{k}{n}}^{\sin x} \left(w - \sin \frac{k}{n}\right)^{N-1} \left|F^{(N)}(w) - F^{(N)}(\sin x)\right| dw \leq$$

$$\frac{1}{(N-1)!} \int_{\sin \frac{k}{n}}^{\sin x} \left(w - \sin \frac{k}{n}\right)^{N-1} \omega_1 \left(F^{(N)}, |w - \sin x|\right) dw \leq$$

$$\frac{1}{(N-1)!} \omega_1 \left(F^{(N)}, \left|\sin x - \sin \frac{k}{n}\right|\right) \int_{\sin \frac{k}{n}}^{\sin x} \left(w - \sin \frac{k}{n}\right)^{N-1} dw \leq$$

$$\frac{1}{(N-1)!} \omega_1 \left(F^{(N)}, \left|x - \frac{k}{n}\right|\right) \frac{\left(\sin x - \sin \frac{k}{n}\right)^N}{N} \leq$$

$$\frac{1}{N!} \omega_1 \left(F^{(N)}, \frac{1}{n^\alpha}\right) \left|x - \frac{k}{n}\right|^N \leq \frac{1}{n^{\alpha N} N!} \omega_1 \left(F^{(N)}, \frac{1}{n^\alpha}\right).$$

We have shown for $\frac{k}{n} \leq x$ that

$$|\rho| \leq \omega_1 \left(F^{(N)}, \frac{1}{n^\alpha}\right) \frac{1}{n^{\alpha N} N!}.$$

So in both cases we got that

$$|\rho| \leq \omega_1\left(F^{(N)}, \frac{1}{n^\alpha}\right)\frac{1}{n^{\alpha N}N!},$$

when $\left|\frac{k}{n} - x\right| \leq \frac{1}{n^\alpha}$.

Also in general ($\frac{k}{n} \geq x$ case)

$$|\rho| \leq \frac{1}{(N-1)!}\left(\int_x^{\frac{k}{n}}\left(\sin\frac{k}{n} - \sin s\right)^{N-1} d\sin s\right) 2\left\|F^{(N)}\right\|_\infty =$$

$$\frac{1}{(N-1)!}\left(\int_{\sin x}^{\sin\frac{k}{n}}\left(\sin\frac{k}{n} - w\right)^{N-1} dw\right) 2\left\|F^{(N)}\right\|_\infty =$$

$$\frac{1}{N!}\left(\sin\frac{k}{n} - \sin x\right)^N 2\left\|F^{(N)}\right\|_\infty \leq \frac{2^{N+1}}{N!}\left\|F^{(N)}\right\|_\infty.$$

Also (case of $\frac{k}{n} \leq x$) we derive

$$|\rho| \leq \frac{1}{(N-1)!}\left(\int_{\frac{k}{n}}^{x}\left(\sin s - \sin\frac{k}{n}\right)^{N-1} d\sin s\right) 2\left\|F^{(N)}\right\|_\infty =$$

$$\frac{1}{(N-1)!}\left(\int_{\sin\frac{k}{n}}^{\sin x}\left(w - \sin\frac{k}{n}\right)^{N-1} dw\right) 2\left\|F^{(N)}\right\|_\infty =$$

$$\frac{1}{(N-1)!}\frac{\left(\sin x - \sin\frac{k}{n}\right)^N}{N} 2\left\|F^{(N)}\right\|_\infty \leq \frac{2^{N+1}}{N!}\left\|F^{(N)}\right\|_\infty.$$

So we proved in general that

$$|\rho| \leq \frac{2^{N+1}}{N!}\left\|F^{(N)}\right\|_\infty.$$

Therefore we have proved

$$|M_n(x)| \leq \sum_{\substack{k=\lceil na\rceil \\ (k:|x-\frac{k}{n}|\leq\frac{1}{n^\alpha})}}^{\lfloor nb\rfloor} \Psi(nx - k)|\rho| + \sum_{\substack{k=\lceil na\rceil \\ (k:|x-\frac{k}{n}|>\frac{1}{n^\alpha})}}^{\lfloor nb\rfloor} \Psi(nx - k)|\rho| \leq$$

$$\left(\frac{\omega_1\left(F^{(N)}, \frac{1}{n^\alpha}\right)}{N!n^{\alpha N}}\right) + \frac{e^4 2^{N+1}}{N!}\left\|F^{(N)}\right\|_\infty e^{-2n^{(1-\alpha)}}.$$

So that

$$|M_n(x)| \leq \frac{\omega_1\left(F^{(N)}, \frac{1}{n^\alpha}\right)}{N!n^{\alpha N}} + \frac{e^4 2^{N+1}}{N!}\left\|F^{(N)}\right\|_\infty e^{-2n^{(1-\alpha)}}.$$

Next we estimate

$$\left| F_n^* \left((\sin \cdot - \sin x)^j , x \right) \right| \le F_n^* \left(|\sin \cdot - \sin x|^j , x \right) \le$$

$$F_n^* \left(|\cdot - x|^j , x \right) = \sum_{k=\lceil na \rceil}^{\lfloor nb \rfloor} \Psi \left(nx - k \right) \left| \frac{k}{n} - x \right|^j \le$$

(work as before)

$$\frac{1}{n^{\alpha j}} + e^4 \left(\pi - 2\varepsilon \right)^j e^{-2n^{(1-\alpha)}} .$$

Therefore

$$\left| F_n^* \left((\sin \cdot - \sin x)^j , x \right) \right| \le \frac{1}{n^{\alpha j}} + e^4 \left(\pi - 2\varepsilon \right)^j e^{-2n^{(1-\alpha)}} ,$$

$j = 1, ..., N$.

The theorem is established.   ∎

We finally present

Theorem 2.20. *Let* $f \in C^N \left([\varepsilon, \pi - \varepsilon] \right)$, $n, N \in \mathbb{N}$, $\varepsilon > 0$ *small*, $x \in [\varepsilon, \pi - \varepsilon]$, $0 < \alpha < 1$. *Then*

1)

$$\left| F_n \left(f, x \right) - \sum_{j=1}^{N} \frac{\left(f \circ \cos^{-1} \right)^{(j)} \left(\cos x \right)}{j!} F_n \left((\cos \cdot - \cos x)^j , x \right) - f \left(x \right) \right| \le$$

$$(2.37)$$

$$(4.1488766) \left[\frac{\omega_1 \left(\left(f \circ \cos^{-1} \right)^{(N)} , \frac{1}{n^\alpha} \right)}{n^{\alpha N} N!} + \right.$$

$$\left. \left(\frac{e^4 2^{N+1} \left\| \left(f \circ \cos^{-1} \right)^{(N)} \right\|_\infty}{N!} \right) e^{-2n^{(1-\alpha)}} \right] ,$$

2)

$$\left| F_n \left(f, x \right) - f \left(x \right) \right| \le (4.1488766) \cdot \qquad (2.38)$$

$$\left\{ \sum_{j=1}^{N} \frac{\left| \left(f \circ \cos^{-1} \right)^{(j)} \left(\cos x \right) \right|}{j!} \left[\frac{1}{n^{\alpha j}} + e^4 \left(\pi - 2\varepsilon \right)^j e^{-2n^{(1-\alpha)}} \right] + \right.$$

$$\left. \left[\frac{\omega_1 \left(\left(f \circ \cos^{-1} \right)^{(N)} , \frac{1}{n^\alpha} \right)}{N! n^{\alpha N}} + \left(\frac{e^4 2^{N+1}}{N!} \left\| \left(f \circ \cos^{-1} \right)^{(N)} \right\|_\infty \right) e^{-2n^{(1-\alpha)}} \right] \right\} ,$$

3) assume further $f^{(j)} \left(x_0 \right) = 0$, $j = 1, ..., N$ *for some* $x_0 \in [\varepsilon, \pi - \varepsilon]$, *it holds*

$$\left| F_n \left(f, x_0 \right) - f \left(x_0 \right) \right| \le (4.1488766) \cdot \qquad (2.39)$$

$$\left[\frac{\omega_1\left((f \circ \cos^{-1})^{(N)}, \frac{1}{n^\alpha}\right)}{n^{\alpha N} N!} + \left(\frac{e^4 2^{N+1} \left\| (f \circ \cos^{-1})^{(N)} \right\|_\infty}{N!} \right) e^{-2n^{(1-\alpha)}} \right].$$

Notice in the last the high speed of convergence of order $n^{-\alpha(N+1)}$.

Proof. Call $F := f \circ \cos^{-1}$ and let $\frac{k}{n}, x \in [\varepsilon, \pi - \varepsilon]$. Then

$$f\left(\frac{k}{n}\right) - f(x) = \sum_{j=1}^{N} \frac{F^{(j)}(\cos x)}{j!} \left(\cos\frac{k}{n} - \cos x\right)^j +$$

$$\frac{1}{(N-1)!} \int_x^{\frac{k}{n}} \left(\cos\frac{k}{n} - \cos s\right)^{N-1} \left(F^{(N)}(\cos s) - F^{(N)}(\cos x)\right) d\cos s.$$

Hence

$$\sum_{k=\lceil na \rceil}^{\lfloor nb \rfloor} f\left(\frac{k}{n}\right) \Psi(nx - k) - f(x) \sum_{k=\lceil na \rceil}^{\lfloor nb \rfloor} \Psi(nx - k) =$$

$$\sum_{j=1}^{N} \frac{F^{(j)}(\cos x)}{j!} \sum_{k=\lceil na \rceil}^{\lfloor nb \rfloor} \Psi(nx - k) \left(\cos\frac{k}{n} - \cos x\right)^j + \frac{1}{(N-1)!} \sum_{k=\lceil na \rceil}^{\lfloor nb \rfloor} \Psi(nx - k) \cdot$$

$$\int_x^{\frac{k}{n}} \left(\cos\frac{k}{n} - \cos s\right)^{N-1} \left(F^{(N)}(\cos s) - F^{(N)}(\cos x)\right) d\cos s.$$

Set here $a = \varepsilon$, $b = \pi - \varepsilon$. Thus

$$F_n^*(f, x) - f(x) \sum_{k=\lceil na \rceil}^{\lfloor nb \rfloor} \Psi(nx - k) =$$

$$\sum_{j=1}^{N} \frac{F^{(j)}(\cos x)}{j!} F_n^* \left((\cos \cdot - \cos x)^j, x\right) + \Theta_n(x),$$

where

$$\Theta_n(x) := \sum_{k=\lceil na \rceil}^{\lfloor nb \rfloor} \Psi(nx - k) \overline{\lambda},$$

with

$$\overline{\lambda} := \frac{1}{(N-1)!} \int_x^{\frac{k}{n}} \left(\cos\frac{k}{n} - \cos s\right)^{N-1} \left(F^{(N)}(\cos s) - F^{(N)}(\cos x)\right) d\cos s =$$

$$\frac{1}{(N-1)!} \int_{\cos x}^{\cos \frac{k}{n}} \left(\cos\frac{k}{n} - w\right)^{N-1} \left(F^{(N)}(w) - F^{(N)}(\cos x)\right) dw.$$

Case of $\left|\frac{k}{n} - x\right| \leq \frac{1}{n^\alpha}$.

i) Subcase of $\frac{k}{n} \geq x$. The function $\cos ine$ is decreasing on $[a, b]$, i.e. $\cos \frac{k}{n} \leq \cos x$.

Then

$$|\overline{\lambda}| \leq \frac{1}{(N-1)!} \int_{\cos \frac{k}{n}}^{\cos x} \left(w - \cos \frac{k}{n}\right)^{N-1} \left|F^{(N)}(w) - F^{(N)}(\cos x)\right| dw \leq$$

$$\frac{1}{(N-1)!} \int_{\cos \frac{k}{n}}^{\cos x} \left(w - \cos \frac{k}{n}\right)^{N-1} \omega_1 \left(F^{(N)}, |w - \cos x|\right) dw \leq$$

$$\omega_1 \left(F^{(N)}, \cos x - \cos \frac{k}{n}\right) \frac{\left(\cos x - \cos \frac{k}{n}\right)^N}{N!} \leq$$

$$\omega_1 \left(F^{(N)}, \left|x - \frac{k}{n}\right|\right) \frac{\left|x - \frac{k}{n}\right|^N}{N!} \leq \omega_1 \left(F^{(N)}, \frac{1}{n^\alpha}\right) \frac{1}{n^{\alpha N} N!}.$$

So if $\frac{k}{n} \geq x$, then

$$|\overline{\lambda}| \leq \omega_1 \left(F^{(N)}, \frac{1}{n^\alpha}\right) \frac{1}{n^{\alpha N} N!}.$$

ii) Subcase of $\frac{k}{n} \leq x$, then $\cos \frac{k}{n} \geq \cos x$. Hence

$$|\overline{\lambda}| \leq \frac{1}{(N-1)!} \int_{\cos x}^{\cos \frac{k}{n}} \left(\cos \frac{k}{n} - w\right)^{N-1} \left|F^{(N)}(w) - F^{(N)}(\cos x)\right| dw \leq$$

$$\frac{1}{(N-1)!} \int_{\cos x}^{\cos \frac{k}{n}} \left(\cos \frac{k}{n} - w\right)^{N-1} \omega_1 \left(F^{(N)}, w - \cos x\right) dw \leq$$

$$\frac{1}{(N-1)!} \omega_1 \left(F^{(N)}, \cos \frac{k}{n} - \cos x\right) \int_{\cos x}^{\cos \frac{k}{n}} \left(\cos \frac{k}{n} - w\right)^{N-1} dw \leq$$

$$\frac{1}{(N-1)!} \omega_1 \left(F^{(N)}, \left|\frac{k}{n} - x\right|\right) \frac{\left(\cos \frac{k}{n} - \cos x\right)^N}{N} \leq$$

$$\frac{1}{N!} \omega_1 \left(F^{(N)}, \left|\frac{k}{n} - x\right|\right) \left|\frac{k}{n} - x\right|^N \leq \frac{1}{N!} \omega_1 \left(F^{(N)}, \frac{1}{n^\alpha}\right) \frac{1}{n^{\alpha N}}.$$

We proved for $\frac{k}{n} \leq x$, that

$$|\overline{\lambda}| \leq \omega_1 \left(F^{(N)}, \frac{1}{n^\alpha}\right) \frac{1}{N! n^{\alpha N}}.$$

So in both cases we got that

$$|\overline{\lambda}| \leq \omega_1 \left(F^{(N)}, \frac{1}{n^\alpha}\right) \frac{1}{n^{\alpha N} N!},$$

when $\left|\frac{k}{n} - x\right| \leq \frac{1}{n^\alpha}$.

Also in general ($\frac{k}{n} \geq x$ case)

$$|\bar{\lambda}| \leq \frac{1}{(N-1)!} \left(\int_{\cos \frac{k}{n}}^{\cos x} \left(w - \cos \frac{k}{n} \right)^{N-1} dw \right) 2 \left\| F^{(N)} \right\|_\infty \leq$$

$$\frac{1}{N!} \left(\cos x - \cos \frac{k}{n} \right)^N 2 \left\| F^{(N)} \right\|_\infty \leq \frac{2^{N+1}}{N!} \left\| F^{(N)} \right\|_\infty.$$

Also (case of $\frac{k}{n} \leq x$) we obtain

$$|\bar{\lambda}| \leq \frac{1}{(N-1)!} \left(\int_{\cos x}^{\cos \frac{k}{n}} \left(\cos \frac{k}{n} - w \right)^{N-1} dw \right) 2 \left\| F^{(N)} \right\|_\infty =$$

$$\frac{1}{N!} \left(\cos \frac{k}{n} - \cos x \right)^N 2 \left\| F^{(N)} \right\|_\infty \leq \frac{2^{N+1}}{N!} \left\| F^{(N)} \right\|_\infty.$$

So we have shown in general that

$$|\bar{\lambda}| \leq \frac{2^{N+1}}{N!} \left\| F^{(N)} \right\|_\infty.$$

Therefore we derive

$$|\Theta_n(x)| \leq \sum_{\substack{k=\lceil na \rceil \\ \left(k : \left| x - \frac{k}{n} \right| \leq \frac{1}{n^\alpha}\right)}}^{\lfloor nb \rfloor} \Psi(nx - k) |\bar{\lambda}| + \sum_{\substack{k=\lceil na \rceil \\ \left(k : \left| x - \frac{k}{n} \right| > \frac{1}{n^\alpha}\right)}}^{\lfloor nb \rfloor} \Psi(nx - k) |\bar{\lambda}| \leq$$

$$\left(\omega_1 \left(F^{(N)}, \frac{1}{n^\alpha} \right) \frac{1}{n^{\alpha N} N!} \right) + e^4 \frac{2^{N+1}}{N!} \left\| F^{(N)} \right\|_\infty e^{-2n^{(1-\alpha)}}.$$

So that

$$|\Theta_n(x)| \leq \frac{\omega_1 \left(F^{(N)}, \frac{1}{n^\alpha} \right)}{n^{\alpha N} N!} + e^4 \frac{2^{N+1}}{N!} \left\| F^{(N)} \right\|_\infty e^{-2n^{(1-\alpha)}}.$$

Next we estimate

$$\left| F_n^* \left((\cos \cdot - \cos x)^j, x \right) \right| \leq F_n^* \left(|\cos \cdot - \cos x|^j, x \right) \leq$$

$$F_n^* \left(|\cdot - x|^j, x \right) = \sum_{k=\lceil na \rceil}^{\lfloor nb \rfloor} \Psi(nx - k) \left| \frac{k}{n} - x \right|^j \leq$$

(work as before)

$$\frac{1}{n^{\alpha j}} + e^4 (\pi - 2\varepsilon)^j e^{-2n^{(1-\alpha)}}.$$

Consequently,

$$\left| F_n^* \left((\cos \cdot - \cos x)^j, x \right) \right| \leq \frac{1}{n^{\alpha j}} + e^4 (\pi - 2\varepsilon)^j e^{-2n^{(1-\alpha)}},$$

$j = 1, ..., N$.

The theorem is proved. ∎

2.4 Complex Neural Network Quantitative Approximations

We make

Remark 2.21. *Let* $X := [a, b]$, $\mathbb{R}$ *and* $f : X \to \mathbb{C}$ *with real and imaginary parts* $f_1, f_2 : f = f_1 + if_2$, $i = \sqrt{-1}$. *Clearly* f *is continuous iff* f_1 *and* f_2 *are continuous.*
  Also it holds

$$f^{(j)}(x) = f_1^{(j)}(x) + if_2^{(j)}(x), \tag{2.40}$$

for all $j = 1, ..., N$, *given that* $f_1, f_2 \in C^N(X)$, $N \in \mathbb{N}$.
  We denote by $C_B(\mathbb{R}, \mathbb{C})$ *the space of continuous and bounded functions* $f : \mathbb{R} \to \mathbb{C}$. *Clearly* f *is bounded, iff both* f_1, f_2 *are bounded from* $\mathbb{R}$ *into* $\mathbb{R}$, *where* $f = f_1 + if_2$.
  Here we define

$$F_n(f, x) := F_n(f_1, x) + iF_n(f_2, x), \tag{2.41}$$

and

$$\overline{F}_n(f, x) := \overline{F}_n(f_1, x) + i\overline{F}_n(f_2, x). \tag{2.42}$$

We see here that

$$|F_n(f, x) - f(x)| \le |F_n(f_1, x) - f_1(x)| + |F_n(f_2, x) - f_2(x)|, \tag{2.43}$$

and

$$\|F_n(f) - f\|_\infty \le \|F_n(f_1) - f_1\|_\infty + \|F_n(f_2) - f_2\|_\infty. \tag{2.44}$$

Similarly we obtain

$$\left|\overline{F}_n(f, x) - f(x)\right| \le \left|\overline{F}_n(f_1, x) - f_1(x)\right| + \left|\overline{F}_n(f_2, x) - f_2(x)\right|, \tag{2.45}$$

and

$$\left\|\overline{F}_n(f) - f\right\|_\infty \le \left\|\overline{F}_n(f_1) - f_1\right\|_\infty + \left\|\overline{F}_n(f_2) - f_2\right\|_\infty. \tag{2.46}$$

  We give

Theorem 2.22. *Let* $f \in C([a, b], \mathbb{C})$, $f = f_1 + if_2$, $0 < \alpha < 1$, $n \in \mathbb{N}$, $x \in [a, b]$. *Then*
  i)

$$|F_n(f, x) - f(x)| \le (4.1488766) \cdot \tag{2.47}$$

$$\left[\left(\omega_1\left(f_1, \frac{1}{n^\alpha}\right) + \omega_1\left(f_2, \frac{1}{n^\alpha}\right)\right) + 2e^4\left(\|f_1\|_\infty + \|f_2\|_\infty\right)e^{-2n^{(1-\alpha)}}\right] =: \Phi_1,$$

and
  ii)

$$\|F_n(f) - f\|_\infty \le \Phi_1. \tag{2.48}$$

Proof. Based on Remark 2.21 and Theorem 2.14.   ∎

We give

Theorem 2.23. *Let* $f \in C_B(\mathbb{R}, \mathbb{C})$, $f = f_1 + if_2$, $0 < \alpha < 1$, $n \in \mathbb{N}$, $x \in \mathbb{R}$. *Then*
   i)

$$\left| \overline{F}_n(f, x) - f(x) \right| \leq \left(\omega_1 \left(f_1, \frac{1}{n^\alpha} \right) + \omega_1 \left(f_2, \frac{1}{n^\alpha} \right) \right) + \tag{2.49}$$

$$2e^4 \left(\|f_1\|_\infty + \|f_2\|_\infty \right) e^{-2n^{(1-\alpha)}} =: \Phi_2,$$

   ii)

$$\left\| \overline{F}_n(f) - f \right\|_\infty \leq \Phi_2. \tag{2.50}$$

Proof. Based on Remark 2.21 and Theorem 2.15.   ∎

Next we present a result of high order complex neural network approximation.

Theorem 2.24. *Let* $f : [a, b] \to \mathbb{C}$, $[a, b] \subset \mathbb{R}$, *such that* $f = f_1 + if_2$. *Suppose* $f_1, f_2 \in C^N([a, b])$, $n, N \in \mathbb{N}$, $0 < \alpha < 1$, $x \in [a, b]$. *Then*
   i)

$$|F_n(f, x) - f(x)| \leq (4.1488766) \cdot \tag{2.51}$$

$$\left\{ \sum_{j=1}^N \frac{\left(\left| f_1^{(j)}(x) \right| + \left| f_2^{(j)}(x) \right| \right)}{j!} \left[\frac{1}{n^{\alpha j}} + e^4 (b-a)^j e^{-2n^{(1-\alpha)}} \right] + \right.$$

$$\left[\frac{\left(\omega_1 \left(f_1^{(N)}, \frac{1}{n^\alpha} \right) + \omega_1 \left(f_2^{(N)}, \frac{1}{n^\alpha} \right) \right)}{n^{\alpha N} N!} + \right.$$

$$\left. \left. \left(\frac{2e^4 \left(\left\| f_1^{(N)} \right\|_\infty + \left\| f_2^{(N)} \right\|_\infty \right) (b-a)^N}{N!} \right) e^{-2n^{(1-\alpha)}} \right] \right\},$$

   ii) assume further $f_1^{(j)}(x_0) = f_2^{(j)}(x_0) = 0$, $j = 1, ..., N$, *for some* $x_0 \in [a, b]$, *it holds*

$$|F_n(f, x_0) - f(x_0)| \leq (4.1488766) \cdot \tag{2.52}$$

$$\left[\frac{\left(\omega_1 \left(f_1^{(N)}, \frac{1}{n^\alpha} \right) + \omega_1 \left(f_2^{(N)}, \frac{1}{n^\alpha} \right) \right)}{n^{\alpha N} N!} + \right.$$

$$\left. \left(\frac{2e^4 \left(\left\| f_1^{(N)} \right\|_\infty + \left\| f_2^{(N)} \right\|_\infty \right) (b-a)^N}{N!} \right) e^{-2n^{(1-\alpha)}} \right],$$

notice here the extremely high rate of convergence at $n^{-(N+1)\alpha}$, *iii)*

$$\|F_n(f) - f\|_\infty \leq (4.1488766) \cdot \tag{2.53}$$

$$\left\{ \sum_{j=1}^{N} \frac{\left(\left\| f_1^{(j)} \right\|_\infty + \left\| f_2^{(j)} \right\|_\infty \right)}{j!} \left[\frac{1}{n^{\alpha j}} + e^4 (b-a)^j e^{-2n^{(1-\alpha)}} \right] + \right.$$

$$\left[\frac{\left(\omega_1 \left(f_1^{(N)}, \frac{1}{n^\alpha} \right) + \omega_1 \left(f_2^{(N)}, \frac{1}{n^\alpha} \right) \right)}{n^{\alpha N} N!} + \right.$$

$$\left. \left. \frac{2e^4 \left(\left\| f_1^{(N)} \right\|_\infty + \left\| f_2^{(N)} \right\|_\infty \right) (b-a)^N}{N!} e^{-2n^{(1-\alpha)}} \right] \right\}.$$

Proof. Based on Remark 2.21 and Theorem 2.16.   ∎

References

[1] Anastassiou, G.A.: Rate of convergence of some neural network operators to the unit-univariate case. J. Math. Anal. Appli. 212, 237–262 (1997)

[2] Anastassiou, G.A.: Quantitative Approximations. Chapman&Hall/CRC, Boca Raton (2001)

[3] Anastassiou, G.A.: Basic Inequalities, Revisited. Mathematica Balkanica, New Series 24, Fasc. 1-2, 59–84 (2010)

[4] Anastassiou, G.A.: Univariate hyperbolic tangent neural network approximation. Mathematics and Computer Modelling 53, 1111–1132 (2011)

[5] Chen, Z., Cao, F.: The approximation operators with sigmoidal functions. Computers and Mathematics with Applications 58, 758–765 (2009)

[6] Haykin, S.: Neural Networks: A Comprehensive Foundation, 2nd edn. Prentice Hall, New York (1998)

[7] Mitchell, T.M.: Machine Learning. WCB-McGraw-Hill, New York (1997)

[8] McCulloch, W., Pitts, W.: A logical calculus of the ideas immanent in nervous activity. Bulletin of Mathematical Biophysics 7, 115–133 (1943)

Chapter 3
Multivariate Sigmoidal Neural Network Quantitative Approximation

Here we present the multivariate quantitative constructive approximation of real and complex valued continuous multivariate functions on a box or $\mathbb{R}^N$, $N \in \mathbb{N}$, by the multivariate quasi-interpolation sigmoidal neural network operators. The "right" operators for the goal are fully and precisely described. This approximation is obtained by establishing multidimensional Jackson type inequalities involving the multivariate modulus of continuity of the engaged function or its high order partial derivatives. The multivariate operators are defined by using a multidimensional density function induced by the logarithmic sigmoidal function. Our approximations are pointwise and uniform. The related feed-forward neural network is with one hidden layer. This chapter is based on [5].

3.1 Introduction

Feed-forward neural networks (FNNs) with one hidden layer, the type of networks we deal with in this chapter, are mathematically expressed in a simplified form as

$$N_n\left(x\right) = \sum_{j=0}^{n} c_j \sigma\left(\langle a_j \cdot x \rangle + b_j\right), \quad x \in \mathbb{R}^s, \quad s \in \mathbb{N},$$

where for $0 \leq j \leq n$, $b_j \in \mathbb{R}$ are the thresholds, $a_j \in \mathbb{R}^s$ are the connection weights, $c_j \in \mathbb{R}$ are the coefficients, $\langle a_j \cdot x \rangle$ is the inner product of a_j and x, and σ is the activation function of the network. In many fundamental network models, the activation function is the sigmoidal function of logistic type.

To achieve our goals the operators here are more elaborate and complex, please see (3.2) and (3.3) for exact definitions.

G.A. Anastassiou: Intelligent Systems: Approximation by ANN, ISRL 19, pp. 67–88.
springerlink.com   © Springer-Verlag Berlin Heidelberg 2011

It is well known that FNNs are universal approximators. Theoretically, any continuous function defined on a compact set can be approximated to any desired degree of accuracy by increasing the number of hidden neurons. It was proved by Cybenko [12] and Funahashi [14], that any continuous function can be approximated on a compact set with uniform topology by a network of the form $N_n(x)$, using any continuous, sigmoidal activation function. Hornik et al. in [16], have shown that any measurable function can be approached with such a network. Furthermore, these authors proved in [17], that any function of the Sobolev spaces can be approached with all derivatives. A variety of density results on FNN approximations to multivariate functions were later established by many authors using different methods, for more or less general situations: [19] by Leshno et al., [23] by Mhaskar and Micchelli, [11] by Chui and Li, [9] by Chen and Chen, [15] by Hahm and Hong, etc.

Usually these results only give theorems about the existence of an approximation. A related and important problem is that of complexity: determining the number of neurons required to guarantee that all functions belonging to a space can be approximated to the prescribed degree of accuracy ϵ.

Barron [6] proves that if the function is supposed to satisfy certain conditions expressed in terms of its Fourier transform, and if each of the neurons evaluates a sigmoidal activation function, then at most $O\left(\epsilon^{-2}\right)$ neurons are needed to achieve the order of approximation ϵ. Some other authors have published similar results on the complexity of FNN approximations: Mhaskar and Micchelli [24], Suzuki [25], Maiorov and Meir [21], Makovoz [22], Ferrari and Stengel [13], Xu and Cao [27], Cao et al. [8], etc.

The author in [1], [2] and [3], see chapters 2-5, was the first to obtain neural network approximations to continuous functions with rates by very specifically defined neural network operators of Cardaliagnet-Euvrard and "Squashing" types, by employing the modulus of continuity of the engaged function or its high order derivative, and producing very tight Jackson type inequalities. He treats there both the univariate and multivariate cases. The defining these operators "bell-shaped" and "squashing" function are assumed to be of compact support. Also in [3] he gives the Nth order asymptotic expansion for the error of weak approximation of these two operators to a special natural class of smooth functions, see chapters 4-5 there.

For this chapter the author is greatly motivated by the important article [10] by Z. Chen and F. Cao, also by [4].

The author here performs multivariate sigmoidal neural network approximations to continuous functions over boxes or over the whole $\mathbb{R}^N$, $N \in \mathbb{N}$, then he extends his results to complex valued multivariate functions. All convergences here are with rates expressed via the multivariate modulus of continuity of the involved function or its high order partial derivatives, and given by very tight multidimensional Jackson type inequalities.

The author here comes up with the right and precisely defined multivariate quasi-interpolation neural network operator related to boxes. The boxes are not necessarily symmetric to the origin. In preparation to prove the main results we prove important properties of the basic multivariate density function defining our operators.

For the approximation theory background we use [20] and [26].

3.2 Background and Auxiliary Results

We consider here the sigmoidal function of logarithmic type

$$s_i(x_i) = \frac{1}{1 + e^{-x_i}}, \quad x_i \in \mathbb{R},\ i = 1, ..., N;\ x := (x_1, ..., x_N) \in \mathbb{R}^N.$$

each has the properties $\lim_{x_i \to +\infty} s_i(x_i) = 1$ and $\lim_{x_i \to -\infty} s_i(x_i) = 0,\ i = 1, ..., N$.

These functions play the role of activation functions in the hidden layer of neural networks, also have applications in biology, demography, etc. ([7, 18]).

As in [10], we consider

$$\Phi_i(x_i) := \frac{1}{2}(s_i(x_i + 1) - s_i(x_i - 1)), \quad x_i \in \mathbb{R},\ i = 1, ..., N.$$

We have the following properties:

i) $\Phi_i(x_i) > 0,\ \forall\ x_i \in \mathbb{R}$,

ii) $\sum_{k_i=-\infty}^{\infty} \Phi_i(x_i - k_i) = 1,\ \forall\ x_i \in \mathbb{R}$,

iii) $\sum_{k_i=-\infty}^{\infty} \Phi_i(nx_i - k_i) = 1,\ \forall\ x_i \in \mathbb{R};\ n \in \mathbb{N}$,

iv) $\int_{-\infty}^{\infty} \Phi_i(x_i)\, dx_i = 1$,

v) Φ_i is a density function,

vi) Φ_i is even: $\Phi_i(-x_i) = \Phi_i(x_i),\ x_i \geq 0$, for $i = 1, ..., N$.

We observe that ([10])

$$\Phi_i(x_i) = \left(\frac{e^2 - 1}{2e^2}\right) \frac{1}{(1 + e^{x_i-1})(1 + e^{-x_i-1})}, \quad i = 1, ..., N.$$

vii) Φ_i is decreasing on $\mathbb{R}_+$, and increasing on $\mathbb{R}_-$, $i = 1, ..., N$.

Let $0 < \beta < 1$, $n \in \mathbb{N}$. Then as in [4] we get

viii)

$$\sum_{\left\{ \substack{k_i = -\infty \\ : |nx_i - k_i| > n^{1-\beta}} \right\}}^{\infty} \Phi_i\left(nx_i - k_i\right) = \sum_{\left\{ \substack{k_i = -\infty \\ : |nx_i - k_i| > n^{1-\beta}} \right\}}^{\infty} \Phi_i\left(|nx_i - k_i|\right)$$

$$\leq 3.1992 e^{-n^{(1-\beta)}}, \quad i = 1, ..., N.$$

Denote by $\lceil \cdot \rceil$ the ceiling of a number, and by $\lfloor \cdot \rfloor$ the integral part of a number. Consider here $x \in \left(\prod_{i=1}^{N} [a_i, b_i] \right) \subset \mathbb{R}^N$, $N \in \mathbb{N}$ such that $\lceil na_i \rceil \leq \lfloor nb_i \rfloor$, $i = 1, ..., N$; $a := (a_1, ..., a_N)$, $b := (b_1, ..., b_N)$.

As in [4] we obtain

ix)

$$0 < \frac{1}{\sum_{k_i=\lceil na_i \rceil}^{\lfloor nb_i \rfloor} \Phi_i\left(nx_i - k_i\right)} < \frac{1}{\Phi_i\left(1\right)} = 5.250312578,$$

$\forall \, x_i \in [a_i, b_i]$, $i = 1, ..., N$.

x) As in [4], we see that

$$\lim_{n \to \infty} \sum_{k_i=\lceil na_i \rceil}^{\lfloor nb_i \rfloor} \Phi_i\left(nx_i - k_i\right) \neq 1,$$

for at least some $x_i \in [a_i, b_i]$, $i = 1, ..., N$.

We will use here

$$\Phi\left(x_1, ..., x_N\right) := \Phi\left(x\right) := \prod_{i=1}^{N} \Phi_i\left(x_i\right), \quad x \in \mathbb{R}^N. \tag{3.1}$$

It has the properties:

(i)' $\Phi\left(x\right) > 0$, $\forall \, x \in \mathbb{R}^N$,

We see that

$$\sum_{k_1=-\infty}^{\infty} \sum_{k_2=-\infty}^{\infty} ... \sum_{k_N=-\infty}^{\infty} \Phi\left(x_1 - k_1, x_2 - k_2, ..., x_N - k_N\right) =$$

$$\sum_{k_1=-\infty}^{\infty} \sum_{k_2=-\infty}^{\infty} ... \sum_{k_N=-\infty}^{\infty} \prod_{i=1}^{N} \Phi_i\left(x_i - k_i\right) = \prod_{i=1}^{N} \left(\sum_{k_i=-\infty}^{\infty} \Phi_i\left(x_i - k_i\right) \right) = 1.$$

That is
(ii)'

$$\sum_{k=-\infty}^{\infty} \Phi\left(x - k\right) := \sum_{k_1=-\infty}^{\infty} \sum_{k_2=-\infty}^{\infty} \cdots \sum_{k_N=-\infty}^{\infty} \Phi\left(x_1 - k_1, \ldots, x_N - k_N\right) = 1,$$

$$k := (k_1, \ldots, k_n), \ \forall \ x \in \mathbb{R}^N.$$

(iii)'

$$\sum_{k=-\infty}^{\infty} \Phi\left(nx - k\right) :=$$

$$\sum_{k_1=-\infty}^{\infty} \sum_{k_2=-\infty}^{\infty} \cdots \sum_{k_N=-\infty}^{\infty} \Phi\left(nx_1 - k_1, \ldots, nx_N - k_N\right) = 1,$$

$$\forall \ x \in \mathbb{R}^N; \ n \in \mathbb{N}.$$

(iv)'

$$\int_{\mathbb{R}^N} \Phi\left(x\right) dx = 1,$$

that is Φ is a multivariate density function.

Here $\|x\|_\infty := \max\{|x_1|, \ldots, |x_N|\}$, $x \in \mathbb{R}^N$, also set $\infty := (\infty, \ldots, \infty)$, $-\infty := (-\infty, \ldots, -\infty)$ upon the multivariate context, and

$$\lceil na \rceil := (\lceil na_1 \rceil, \ldots, \lceil na_N \rceil),$$
$$\lfloor nb \rfloor := (\lfloor nb_1 \rfloor, \ldots, \lfloor nb_N \rfloor).$$

For $0 < \beta < 1$ and $n \in \mathbb{N}$, fixed $x \in \mathbb{R}^N$, have that

$$\sum_{k=\lceil na \rceil}^{\lfloor nb \rfloor} \Phi\left(nx - k\right) =$$

$$\sum_{\substack{k = \lceil na \rceil \\ \left\| \frac{k}{n} - x \right\|_\infty \le \frac{1}{n^\beta}}}^{\lfloor nb \rfloor} \Phi\left(nx - k\right) + \sum_{\substack{k = \lceil na \rceil \\ \left\| \frac{k}{n} - x \right\|_\infty > \frac{1}{n^\beta}}}^{\lfloor nb \rfloor} \Phi\left(nx - k\right).$$

In the last two sums the counting is over disjoint vector of k's, because the condition $\left\| \frac{k}{n} - x \right\|_\infty > \frac{1}{n^\beta}$ implies that there exists at least one $\left| \frac{k_r}{n} - x_r \right| > \frac{1}{n^\beta}$, $r \in \{1, \ldots, N\}$.

We treat

$$\sum_{\substack{k = \lceil na \rceil \\ \left\| \frac{k}{n} - x \right\|_\infty > \frac{1}{n^\beta}}}^{\lfloor nb \rfloor} \Phi\left(nx - k\right) = \prod_{i=1}^{N} \left(\sum_{\substack{k_i = \lceil na_i \rceil \\ \left\| \frac{k}{n} - x \right\|_\infty > \frac{1}{n^\beta}}}^{\lfloor nb_i \rfloor} \Phi_i\left(nx_i - k_i\right) \right)$$

$$\leq \left(\prod_{\substack{i=1 \\ i\neq r}}^{N} \left(\sum_{k_i=-\infty}^{\infty} \Phi_i \left(n x_i - k_i \right) \right) \right) \cdot \left(\sum_{\substack{\left\{ \begin{array}{l} k_r = \lceil n a_r \rceil \\ \left| \frac{k_r}{n} - x_r \right| > \frac{1}{n^\beta} \end{array} \right.}}^{\lfloor n b_r \rfloor} \Phi_r \left(n x_r - k_r \right) \right)$$

$$= \left(\sum_{\substack{\left\{ \begin{array}{l} k_r = \lceil n a_r \rceil \\ \left| \frac{k_r}{n} - x_r \right| > \frac{1}{n^\beta} \end{array} \right.}}^{\lfloor n b_r \rfloor} \Phi_r \left(n x_r - k_r \right) \right)$$

$$\leq \sum_{\substack{\left\{ \begin{array}{l} k_r = -\infty \\ \left| \frac{k_r}{n} - x_r \right| > \frac{1}{n^\beta} \end{array} \right.}}^{\infty} \Phi_r \left(n x_r - k_r \right) \overset{\text{(by (viii))}}{\leq} 3.1992 e^{-n^{(1-\beta)}}.$$

We have established that

(v)'

$$\sum_{\substack{\left\{ \begin{array}{l} k = \lceil n a \rceil \\ \left\| \frac{k}{n} - x \right\|_\infty > \frac{1}{n^\beta} \end{array} \right.}}^{\lfloor n b \rfloor} \Phi \left(n x - k \right) \leq 3.1992 e^{-n^{(1-\beta)}},$$

$0 < \beta < 1$, $n \in \mathbb{N}$, $x \in \left(\prod_{i=1}^{N} [a_i, b_i] \right)$.

By (ix) clearly we obtain

$$0 < \frac{1}{\sum_{k=\lceil na \rceil}^{\lfloor nb \rfloor} \Phi \left(nx - k \right)} = \frac{1}{\prod_{i=1}^{N} \left(\sum_{k_i=\lceil na_i \rceil}^{\lfloor nb_i \rfloor} \Phi_i \left(n x_i - k_i \right) \right)}$$

$$< \frac{1}{\prod_{i=1}^{N} \Phi_i \left(1 \right)} = \left(5.250312578 \right)^{N}.$$

That is,

(vi)' it holds

$$0 < \frac{1}{\sum_{k=\lceil na \rceil}^{\lfloor nb \rfloor} \Phi \left(nx - k \right)} < \left(5.250312578 \right)^{N},$$

$\forall \ x \in \left(\prod_{i=1}^{N} [a_i, b_i] \right)$, $n \in \mathbb{N}$.

It is also obvious that

(vii)'

$$\sum_{\substack{k=-\infty \\ \left\|\frac{k}{n}-x\right\|_\infty > \frac{1}{n^\beta}}}^{\infty} \Phi\left(nx-k\right) \le 3.1992 e^{-n^{(1-\beta)}},$$

$0 < \beta < 1$, $n \in \mathbb{N}$, $x \in \mathbb{R}^N$.

By (x) we obviously observe that

(viii)'

$$\lim_{n\to\infty} \sum_{k=\lceil na \rceil}^{\lfloor nb \rfloor} \Phi\left(nx-k\right) \ne 1$$

for at least some $x \in \left(\prod_{i=1}^{N} [a_i, b_i]\right)$.

Let $f \in C\left(\prod_{i=1}^{N} [a_i, b_i]\right)$ and $n \in \mathbb{N}$ such that $\lceil na_i \rceil \le \lfloor nb_i \rfloor$, $i = 1, ..., N$. We introduce and define the multivariate positive linear neural network operator $(x := (x_1, ..., x_N) \in \left(\prod_{i=1}^{N} [a_i, b_i]\right))$

$$G_n\left(f, x_1, ..., x_N\right) := G_n\left(f, x\right) := \frac{\sum_{k=\lceil na \rceil}^{\lfloor nb \rfloor} f\left(\frac{k}{n}\right) \Phi\left(nx-k\right)}{\sum_{k=\lceil na \rceil}^{\lfloor nb \rfloor} \Phi\left(nx-k\right)} \tag{3.2}$$

$$:= \frac{\sum_{k_1=\lceil na_1 \rceil}^{\lfloor nb_1 \rfloor} \sum_{k_2=\lceil na_2 \rceil}^{\lfloor nb_2 \rfloor} \cdots \sum_{k_N=\lceil na_N \rceil}^{\lfloor nb_N \rfloor} f\left(\frac{k_1}{n}, ..., \frac{k_N}{n}\right) \left(\prod_{i=1}^{N} \Phi_i\left(nx_i-k_i\right)\right)}{\prod_{i=1}^{N} \left(\sum_{k_i=\lceil na_i \rceil}^{\lfloor nb_i \rfloor} \Phi_i\left(nx_i-k_i\right)\right)}.$$

For large enough n we always obtain $\lceil na_i \rceil \le \lfloor nb_i \rfloor$, $i = 1, ..., N$. Also $a_i \le \frac{k_i}{n} \le b_i$, iff $\lceil na_i \rceil \le k_i \le \lfloor nb_i \rfloor$, $i = 1, ..., N$.

We study here the pointwise and uniform convergence of $G_n\left(f\right)$ to f with rates.

For convenience we call

$$G_n^*\left(f, x\right) := \sum_{k=\lceil na \rceil}^{\lfloor nb \rfloor} f\left(\frac{k}{n}\right) \Phi\left(nx-k\right) \tag{3.3}$$

$$:= \sum_{k_1=\lceil na_1 \rceil}^{\lfloor nb_1 \rfloor} \sum_{k_2=\lceil na_2 \rceil}^{\lfloor nb_2 \rfloor} \cdots \sum_{k_N=\lceil na_N \rceil}^{\lfloor nb_N \rfloor} f\left(\frac{k_1}{n}, ..., \frac{k_N}{n}\right) \left(\prod_{i=1}^{N} \Phi_i\left(nx_i-k_i\right)\right),$$

$\forall\, x \in \left(\prod_{i=1}^{N} [a_i, b_i]\right)$.

That is

$$G_n\left(f,x\right) := \frac{G_n^*\left(f,x\right)}{\sum_{k=\lceil na \rceil}^{\lfloor nb \rfloor} \Phi\left(nx - k\right)}, \quad \forall x \in \left(\prod_{i=1}^{N} [a_i, b_i]\right), \; n \in \mathbb{N}. \qquad (3.4)$$

Therefore

$$G_n\left(f,x\right) - f\left(x\right) = \frac{G_n^*\left(f,x\right) - f\left(x\right) \sum_{k=\lceil na \rceil}^{\lfloor nb \rfloor} \Phi\left(nx - k\right)}{\sum_{k=\lceil na \rceil}^{\lfloor nb \rfloor} \Phi\left(nx - k\right)}. \qquad (3.5)$$

Consequently we derive

$$\left|G_n\left(f,x\right) - f\left(x\right)\right| \le (5.250312578)^N \left|G_n^*\left(f,x\right) - f\left(x\right) \sum_{k=\lceil na \rceil}^{\lfloor nb \rfloor} \Phi\left(nx - k\right)\right|, \qquad (3.6)$$

$\forall \, x \in \left(\prod_{i=1}^{N} [a_i, b_i]\right).$

We will estimate the right hand side of (3.6).

For that we need, for $f \in C\left(\prod_{i=1}^{N} [a_i, b_i]\right)$ the first multivariate modulus of continuity

$$\omega_1\left(f,h\right) := \sup_{\substack{x,y \in \left(\prod_{i=1}^{N} [a_i, b_i]\right) \\ \|x - y\|_\infty \le h}} \left|f\left(x\right) - f\left(y\right)\right|, \quad h > 0. \qquad (3.7)$$

Similarly it is defined for $f \in C_B\left(\mathbb{R}^N\right)$ (continuous and bounded functions on $\mathbb{R}^N$). We have that $\lim_{h \to 0} \omega_1\left(f,h\right) = 0$.

When $f \in C_B\left(\mathbb{R}^N\right)$ we define

$$\overline{G}_n\left(f,x\right) := \overline{G}_n\left(f,x_1, ..., x_N\right) := \sum_{k=-\infty}^{\infty} f\left(\frac{k}{n}\right) \Phi\left(nx - k\right) \qquad (3.8)$$

$$:= \sum_{k_1=-\infty}^{\infty} \sum_{k_2=-\infty}^{\infty} \cdots \sum_{k_N=-\infty}^{\infty} f\left(\frac{k_1}{n}, \frac{k_2}{n}, ..., \frac{k_N}{n}\right) \left(\prod_{i=1}^{N} \Phi_i\left(nx_i - k_i\right)\right),$$

$n \in \mathbb{N}$, $\forall \, x \in \mathbb{R}^N$, $N \ge 1$, the multivariate quasi-interpolation neural network operator.

Notice here that for large enough $n \in \mathbb{N}$ we get that

$$e^{-n^{(1-\beta)}} < n^{-\beta j}, \quad j = 1, ..., m \in \mathbb{N}, \; 0 < \beta < 1. \qquad (3.9)$$

Thus be given fixed $A, B > 0$, for the linear combination $\left(An^{-\beta j} + Be^{-n^{(1-\beta)}}\right)$ the (dominant) rate of convergence to zero is $n^{-\beta j}$. The closer β is to 1 we get faster and better rate of convergence to zero.

Let $f \in C^m\left(\prod_{i=1}^{N}[a_i, b_i]\right)$, $m, N \in \mathbb{N}$. Here f_α denotes a partial derivative of f, $\alpha := (\alpha_1, ..., \alpha_N)$, $\alpha_i \in \mathbb{Z}^+$, $i = 1, ..., N$, and $|\alpha| := \sum_{i=1}^{N} \alpha_i = l$, where $l = 0, 1, ..., m$. We write also $f_\alpha := \frac{\partial^\alpha f}{\partial x^\alpha}$ and we say it is of order l.

We denote

$$\omega_{1,m}^{\max}(f_\alpha, h) := \max_{\alpha:|\alpha|=m} \omega_1(f_\alpha, h). \tag{3.10}$$

Call also

$$\|f_\alpha\|_{\infty,m}^{\max} := \max_{|\alpha|=m}\{\|f_\alpha\|_\infty\}, \tag{3.11}$$

$\|\cdot\|_\infty$ is the supremum norm.

3.3 Real Multivariate Neural Network Quantitative Approximations

Here we present a series of multivariate neural network approximations to a function given with rates.

We first present

Theorem 3.1. Let $f \in C\left(\prod_{i=1}^{N}[a_i, b_i]\right)$, $0 < \beta < 1$, $x \in \left(\prod_{i=1}^{N}[a_i, b_i]\right)$, $n, N \in \mathbb{N}$. Then

i)

$$|G_n(f, x) - f(x)| \leq (5.250312578)^N \cdot$$

$$\left\{\omega_1\left(f, \frac{1}{n^\beta}\right) + (6.3984)\|f\|_\infty e^{-n^{(1-\beta)}}\right\} =: \lambda_1, \tag{3.12}$$

ii)

$$\|G_n(f) - f\|_\infty \leq \lambda_1. \tag{3.13}$$

Proof. We observe that

$$\Delta(x) := G_n^*(f, x) - f(x) \sum_{k=\lceil na \rceil}^{\lfloor nb \rfloor} \Phi(nx - k) =$$

$$\sum_{k=\lceil na \rceil}^{\lfloor nb \rfloor} f\left(\frac{k}{n}\right)\Phi(nx - k) - \sum_{k=\lceil na \rceil}^{\lfloor nb \rfloor} f(x)\Phi(nx - k) =$$

$$\sum_{k=\lceil na \rceil}^{\lfloor nb \rfloor}\left(f\left(\frac{k}{n}\right) - f(x)\right)\Phi(nx - k).$$

So that

$$|\Delta(x)| \leq \sum_{k=\lceil na \rceil}^{\lfloor nb \rfloor} \left| f\left(\frac{k}{n}\right) - f(x) \right| \Phi(nx - k) =$$

$$\sum_{\substack{k = \lceil na \rceil \\ \left\| \frac{k}{n} - x \right\|_\infty \leq \frac{1}{n^\beta}}} \left| f\left(\frac{k}{n}\right) - f(x) \right| \Phi(nx - k) +$$

$$\sum_{\substack{k = \lceil na \rceil \\ \left\| \frac{k}{n} - x \right\|_\infty > \frac{1}{n^\beta}}}^{\lfloor nb \rfloor} \left| f\left(\frac{k}{n}\right) - f(x) \right| \Phi(nx - k) \leq$$

$$\omega_1\left(f, \frac{1}{n^\beta}\right) + 2\|f\|_\infty \sum_{\substack{k = \lceil na \rceil \\ \left\| \frac{k}{n} - x \right\|_\infty > \frac{1}{n^\beta}}}^{\lfloor nb \rfloor} \Phi(nx - k) \leq$$

$$\omega_1\left(f, \frac{1}{n^\beta}\right) + (6.3984)\|f\|_\infty \, e^{-n^{(1-\beta)}}.$$

So that

$$|\Delta| \leq \omega_1\left(f, \frac{1}{n^\beta}\right) + (6.3984)\|f\|_\infty \, e^{-n^{(1-\beta)}}.$$

Now using (3.6) we prove claim.   ∎

Next we give

Theorem 3.2. *Let* $f \in C_B\left(\mathbb{R}^N\right)$, $0 < \beta < 1$, $x \in \mathbb{R}^N$, $n, N \in \mathbb{N}$. *Then*
  i)

$$\left| \overline{G}_n(f, x) - f(x) \right| \leq \omega_1\left(f, \frac{1}{n^\beta}\right) + (6.3984)\|f\|_\infty \, e^{-n^{(1-\beta)}} =: \lambda_2, \quad (3.14)$$

  ii)

$$\left\| \overline{G}_n(f) - f \right\|_\infty \leq \lambda_2. \quad (3.15)$$

Proof. We have

$$\overline{G}_n(f, x) := \sum_{k=-\infty}^{\infty} f\left(\frac{k}{n}\right) \Phi(nx - k).$$

Hence

$$E_n(x) := \overline{G}_n(f, x) - f(x) = \sum_{k=-\infty}^{\infty} f\left(\frac{k}{n}\right) \Phi(nx - k) - f(x) \sum_{k=-\infty}^{\infty} \Phi(nx - k) =$$

$$\sum_{k=-\infty}^{\infty} \left(f\left(\frac{k}{n}\right) - f(x) \right) \Phi(nx - k).$$

Thus

$$|E_n(x)| \leq \sum_{k=-\infty}^{\infty} \left| f\left(\frac{k}{n}\right) - f(x) \right| \Phi(nx - k) =$$

$$\sum_{\substack{k=-\infty \\ \left\| \frac{k}{n} - x \right\|_\infty \leq \frac{1}{n^\beta}}}^{\infty} \left| f\left(\frac{k}{n}\right) - f(x) \right| \Phi(nx - k) +$$

$$\sum_{\substack{k=-\infty \\ \left\| \frac{k}{n} - x \right\|_\infty > \frac{1}{n^\beta}}}^{\infty} \left| f\left(\frac{k}{n}\right) - f(x) \right| \Phi(nx - k) \leq$$

$$\omega_1\left(f, \frac{1}{n^\beta}\right) + 2\|f\|_\infty \sum_{\substack{k=-\infty \\ \left\| \frac{k}{n} - x \right\|_\infty > \frac{1}{n^\beta}}}^{\infty} \Phi(nx - k) \leq$$

$$\omega_1\left(f, \frac{1}{n^\beta}\right) + (6.3984)\|f\|_\infty \, e^{-n^{(1-\beta)}}.$$

Consequently,

$$|E_n(x)| \leq \omega_1\left(f, \frac{1}{n^\beta}\right) + (6.3984)\|f\|_\infty \, e^{-n^{(1-\beta)}},$$

proving the claim. ∎

In the next we discuss high order of approximation by using the smoothness of f.

We present

Theorem 3.3. *Let* $f \in C^m\left(\prod_{i=1}^N [a_i, b_i]\right)$, $0 < \beta < 1$, $n, m, N \in \mathbb{N}$, $x \in \left(\prod_{i=1}^N [a_i, b_i]\right)$. *Then*
i)

$$\left| G_n(f, x) - f(x) - \sum_{j=1}^m \left(\sum_{|\alpha|=j} \left(\frac{f_\alpha(x)}{\prod_{i=1}^N \alpha_i!} \right) G_n\left(\prod_{i=1}^N (\cdot - x_i)^{\alpha_i}, x \right) \right) \right| \leq$$

$$(5.250312578)^N \cdot \left\{ \frac{N^m}{m! n^{m\beta}} \omega_{1,m}^{\max}\left(f_\alpha, \frac{1}{n^\beta}\right) + \left(\frac{(6.3984)\|b - a\|_\infty^m \|f_\alpha\|_{\infty,m}^{\max} N^m}{m!} \right) e^{-n^{(1-\beta)}} \right\},$$

(3.16)

ii)
$$|G_n(f, x) - f(x)| \leq (5.250312578)^N \cdot$$
(3.17)

$$\left\{ \sum_{j=1}^{m} \left(\sum_{|\alpha|=j} \left(\frac{|f_\alpha(x)|}{\prod_{i=1}^{N} \alpha_i!} \right) \left[\frac{1}{n^{\beta j}} + \left(\prod_{i=1}^{N} (b_i - a_i)^{\alpha_i} \right) \cdot (3.1992) e^{-n^{(1-\beta)}} \right] \right) + \right.$$
$$\left. \frac{N^m}{m! n^{m\beta}} \omega_{1,m}^{\max} \left(f_\alpha, \frac{1}{n^\beta} \right) + \left(\frac{(6.3984) \|b-a\|_\infty^m \|f_\alpha\|_{\infty,m}^{\max} N^m}{m!} \right) e^{-n^{(1-\beta)}} \right\},$$

iii)
$$\|G_n(f) - f\|_\infty \leq (5.250312578)^N \cdot$$
(3.18)

$$\left\{ \sum_{j=1}^{N} \left(\sum_{|\alpha|=j} \left(\frac{\|f_\alpha\|_\infty}{\prod_{i=1}^{N} \alpha_i!} \right) \left[\frac{1}{n^{\beta j}} + \left(\prod_{i=1}^{N} (b_i - a_i)^{\alpha_i} \right) (3.1992) e^{-n^{(1-\beta)}} \right] \right) + \right.$$
$$\left. \frac{N^m}{m! n^{m\beta}} \omega_{1,m}^{\max} \left(f_\alpha, \frac{1}{n^\beta} \right) + \left(\frac{(6.3984) \|b-a\|_\infty^m \|f_\alpha\|_{\infty,m}^{\max} N^m}{m!} \right) e^{-n^{(1-\beta)}} \right\},$$

iv) Suppose $f_\alpha(x_0) = 0$, *for all* $\alpha : |\alpha| = 1, ..., m$; $x_0 \in \left(\prod_{i=1}^{N} [a_i, b_i] \right)$.
Then
$$|G_n(f, x_0) - f(x_0)| \leq (5.250312578)^N \cdot$$
(3.19)

$$\left\{ \frac{N^m}{m! n^{m\beta}} \omega_1^{\max} \left(f_\alpha, \frac{1}{n^\beta} \right) + \left(\frac{(6.3984) \|b-a\|_\infty^m \|f_\alpha\|_{\infty,m}^{\max} N^m}{m!} \right) e^{-n^{(1-\beta)}} \right\},$$

notice in the last the extremely high rate of convergence at $n^{-\beta(m+1)}$.

Proof. Consider $g_z(t) := f(x_0 + t(z - x_0))$, $t \geq 0$; $x_0, z \in \prod_{i=1}^{N} [a_i, b_i]$.
Then

$$g_z^{(j)}(t) = \left[\left(\sum_{i=1}^{N} (z_i - x_{0i}) \frac{\partial}{\partial x_i} \right)^j f \right] (x_{01} + t(z_1 - x_{01}), ..., x_{0N} + t(z_N - x_{0N})),$$

for all $j = 0, 1, ..., m$.
We have the multivariate Taylor's formula

$$f(z_1, ..., z_N) = g_z(1) =$$

$$\sum_{j=0}^{m} \frac{g_z^{(j)}(0)}{j!} + \frac{1}{(m-1)!} \int_0^1 (1-\theta)^{m-1} \left(g_z^{(m)}(\theta) - g_z^{(m)}(0) \right) d\theta.$$

Notice $g_z(0) = f(x_0)$. Also for $j = 0, 1, ..., m$, we have

$$g_z^{(j)}(0) = \sum_{\substack{\alpha:=(\alpha_1,...,\alpha_N),\ \alpha_i \in \mathbb{Z}^+, \\ i=1,...,N,\ |\alpha|:=\sum_{i=1}^{N} \alpha_i = j}} \left(\frac{j!}{\prod_{i=1}^{N} \alpha_i!} \right) \left(\prod_{i=1}^{N} (z_i - x_{0i})^{\alpha_i} \right) f_\alpha(x_0).$$

Furthermore

$$g_z^{(m)}(\theta) =$$

$$\sum_{\substack{\alpha:=(\alpha_1,...,\alpha_N),\ \alpha_i \in \mathbb{Z}^+,\\ i=1,...,N,\ |\alpha|:=\sum_{i=1}^N \alpha_i = m}} \left(\frac{m!}{\prod_{i=1}^N \alpha_i!}\right) \left(\prod_{i=1}^N (z_i - x_{0i})^{\alpha_i}\right) f_\alpha(x_0 + \theta(z - x_0)),$$

$0 \le \theta \le 1$.

So we treat $f \in C^m\left(\prod_{i=1}^N [a_i, b_i]\right)$.

Thus, we have for $\frac{k}{n}, x \in \left(\prod_{i=1}^N [a_i, b_i]\right)$ that

$$f\left(\frac{k_1}{n}, ..., \frac{k_N}{n}\right) - f(x) =$$

$$\sum_{j=1}^m \sum_{\substack{\alpha:=(\alpha_1,...,\alpha_N),\ \alpha_i \in \mathbb{Z}^+,\\ i=1,...,N,\ |\alpha|:=\sum_{i=1}^N \alpha_i = j}} \left(\frac{1}{\prod_{i=1}^N \alpha_i!}\right) \left(\prod_{i=1}^N \left(\frac{k_i}{n} - x_i\right)^{\alpha_i}\right) f_\alpha(x) + R,$$

where

$$R := m \int_0^1 (1-\theta)^{m-1} \sum_{\substack{\alpha:=(\alpha_1,...,\alpha_N),\ \alpha_i \in \mathbb{Z}^+,\\ i=1,...,N,\ |\alpha|:=\sum_{i=1}^N \alpha_i = m}} \left(\frac{1}{\prod_{i=1}^N \alpha_i!}\right) \left(\prod_{i=1}^N \left(\frac{k_i}{n} - x_i\right)^{\alpha_i}\right)$$

$$\cdot \left[f_\alpha\left(x + \theta\left(\frac{k}{n} - x\right)\right) - f_\alpha(x)\right] d\theta.$$

We observe that

$$|R| \le m \int_0^1 (1-\theta)^{m-1} \sum_{|\alpha|=m} \left(\frac{1}{\prod_{i=1}^N \alpha_i!}\right) \left(\prod_{i=1}^N \left|\frac{k_i}{n} - x_i\right|^{\alpha_i}\right) \cdot$$

$$\left|f_\alpha\left(x + \theta\left(\frac{k}{n} - x\right)\right) - f_\alpha(x)\right| d\theta \le m \int_0^1 (1-\theta)^{m-1} \cdot$$

$$\sum_{|\alpha|=m} \left(\frac{1}{\prod_{i=1}^N \alpha_i!}\right) \left(\prod_{i=1}^N \left|\frac{k_i}{n} - x_i\right|^{\alpha_i}\right) \omega_1\left(f_\alpha, \theta\left\|\frac{k}{n} - x\right\|_\infty\right) d\theta \le (*).$$

Notice here that

$$\left\|\frac{k}{n} - x\right\|_\infty \le \frac{1}{n^\beta} \Leftrightarrow \left|\frac{k_i}{n} - x_i\right| \le \frac{1}{n^\beta}, \quad i = 1, ..., N.$$

We further see that

$$(*) \le m \cdot \omega_{1,m}^{\max}\left(f_\alpha, \frac{1}{n^\beta}\right) \int_0^1 (1-\theta)^{m-1} \sum_{|\alpha|=m} \left(\frac{1}{\prod_{i=1}^N \alpha_i!}\right) \left(\prod_{i=1}^N \left(\frac{1}{n^\beta}\right)^{\alpha_i}\right) d\theta =$$

$$\left(\frac{\omega_{1,m}^{\max} \left(f_\alpha, \frac{1}{n^\beta} \right)}{(m!) \, n^{m\beta}} \right) \left(\sum_{|\alpha|=m} \frac{m!}{\prod_{i=1}^{N} \alpha_i!} \right) = \left(\frac{\omega_{1,m}^{\max} \left(f_\alpha, \frac{1}{n^\beta} \right)}{(m!) \, n^{m\beta}} \right) N^m.$$

Conclusion: When $\left\| \frac{k}{n} - x \right\|_\infty \leq \frac{1}{n^\beta}$, we proved that

$$|R| \leq \left(\frac{N^m}{m! n^{m\beta}} \right) \omega_{1,m}^{\max} \left(f_\alpha, \frac{1}{n^\beta} \right).$$

In general we notice that

$$|R| \leq m \int_0^1 (1-\theta)^{m-1} \left(\sum_{|\alpha|=m} \left(\frac{1}{\prod_{i=1}^{N} \alpha_i!} \right) \left(\prod_{i=1}^{N} (b_i - a_i)^{\alpha_i} \right) 2 \, \|f_\alpha\|_\infty \right) d\theta =$$

$$2 \sum_{|\alpha|=m} \frac{1}{\prod_{i=1}^{N} \alpha_i!} \left(\prod_{i=1}^{N} (b_i - a_i)^{\alpha_i} \right) \|f_\alpha\|_\infty \leq$$

$$\left(\frac{2 \, \|b - a\|_\infty^m \, \|f_\alpha\|_{\infty,m}^{\max}}{m!} \right) \left(\sum_{|\alpha|=m} \frac{m!}{\prod_{i=1}^{N} \alpha_i!} \right) = \frac{2 \, \|b - a\|_\infty^m \, \|f_\alpha\|_{\infty,m}^{\max} \, N^m}{m!}.$$

We proved in general that

$$|R| \leq \frac{2 \, \|b - a\|_\infty^m \, \|f_\alpha\|_{\infty,m}^{\max} \, N^m}{m!} := \lambda_3.$$

Next we observe that

$$U_n := \sum_{k=\lceil na \rceil}^{\lfloor nb \rfloor} \Phi (nx - k) \, R =$$

$$\sum_{\substack{k=\lceil na \rceil \\ :\left\| \frac{k}{n} - x \right\|_\infty \leq \frac{1}{n^\beta}}}^{\lfloor nb \rfloor} \Phi (nx - k) \, R + \sum_{\substack{k=\lceil na \rceil \\ :\left\| \frac{k}{n} - x \right\|_\infty > \frac{1}{n^\beta}}}^{\lfloor nb \rfloor} \Phi (nx - k) \, R.$$

Consequently

$$|U_n| \leq \left(\sum_{\substack{k=\lceil na \rceil \\ :\left\| \frac{k}{n} - x \right\|_\infty \leq \frac{1}{n^\beta}}}^{\lfloor nb \rfloor} \Phi (nx - k) \right) \frac{N^m}{m! n^{m\beta}} \omega_{1,m}^{\max} \left(f_\alpha, \frac{1}{n^\beta} \right) + (3.1992) \, \lambda_3 e^{-n^{(1-\beta)}}$$

$$\leq \frac{N^m}{m! n^{m\beta}} \omega_{1,m}^{\max} \left(f_\alpha, \frac{1}{n^\beta} \right) + (3.1992) \, \lambda_3 e^{-n^{(1-\beta)}}.$$

We have established that

$$|U_n| \le \frac{N^m}{m!n^{m\beta}} \omega_{1,m}^{\max}\left(f_\alpha, \frac{1}{n^\beta}\right) + \left(\frac{(6.3984)\,\|b-a\|_\infty^m\,\|f_\alpha\|_{\infty,m}^{\max}\,N^m}{m!}\right)e^{-n^{(1-\beta)}}.$$

We see that

$$\sum_{k=\lceil na\rceil}^{\lfloor nb\rfloor} f\left(\frac{k}{n}\right)\Phi(nx-k) - f(x)\sum_{k=\lceil na\rceil}^{\lfloor nb\rfloor}\Phi(nx-k) =$$

$$\sum_{j=1}^{m}\left(\sum_{|\alpha|=j}\left(\frac{f_\alpha(x)}{\prod_{i=1}^N \alpha_i!}\right)\left(\sum_{k=\lceil na\rceil}^{\lfloor nb\rfloor}\Phi(nx-k)\left(\prod_{i=1}^N\left(\frac{k_i}{n}-x_i\right)^{\alpha_i}\right)\right)\right)$$

$$+\sum_{k=\lceil na\rceil}^{\lfloor nb\rfloor}\Phi(nx-k)\,R.$$

The last says that

$$G_n^*(f,x) - f(x)\left(\sum_{k=\lceil na\rceil}^{\lfloor nb\rfloor}\Phi(nx-k)\right) -$$

$$\sum_{j=1}^{m}\left(\sum_{|\alpha|=j}\left(\frac{f_\alpha(x)}{\prod_{i=1}^N \alpha_i!}\right)G_n^*\left(\prod_{i=1}^N(\cdot-x_i)^{\alpha_i},x\right)\right) = U_n.$$

Clearly G_n^* is a positive linear operator.

Thus (here $\alpha_i \in \mathbb{Z}^+ : |\alpha| = \sum_{i=1}^N \alpha_i = j$)

$$\left|G_n^*\left(\prod_{i=1}^N(\cdot-x_i)^{\alpha_i},x\right)\right| \le G_n^*\left(\prod_{i=1}^N|\cdot-x_i|^{\alpha_i},x\right) =$$

$$\sum_{k=\lceil na\rceil}^{\lfloor nb\rfloor}\left(\prod_{i=1}^N\left|\frac{k_i}{n}-x_i\right|^{\alpha_i}\right)\Phi(nx-k) =$$

$$\sum_{\substack{k=\lceil na\rceil \\ :\left\|\frac{k}{n}-x\right\|_\infty\le\frac{1}{n^\beta}}}^{\lfloor nb\rfloor}\left(\prod_{i=1}^N\left|\frac{k_i}{n}-x_i\right|^{\alpha_i}\right)\Phi(nx-k) +$$

$$\sum_{\substack{k=\lceil na\rceil \\ :\left\|\frac{k}{n}-x\right\|_\infty>\frac{1}{n^\beta}}}^{\lfloor nb\rfloor}\left(\prod_{i=1}^N\left|\frac{k_i}{n}-x_i\right|^{\alpha_i}\right)\Phi(nx-k) \le$$

$$\frac{1}{n^{\beta j}} + \prod_{i=1}^{N} (b_i - a_i)^{\alpha_i} \left(\sum_{\substack{k=\lceil na \rceil \\ :\left\| \frac{k}{n} - x \right\|_\infty > \frac{1}{n^\beta}}}^{\lfloor nb \rfloor} \Phi (nx - k) \right) \le$$

$$\frac{1}{n^{\beta j}} + \left(\prod_{i=1}^{N} (b_i - a_i)^{\alpha_i} \right) (3.1992) \, e^{-n^{(1-\beta)}}.$$

So we have proved that

$$\left| G_n^* \left(\prod_{i=1}^{N} (\cdot - x_i)^{\alpha_i} , x \right) \right| \le \frac{1}{n^{\beta j}} + \left(\prod_{i=1}^{N} (b_i - a_i)^{\alpha_i} \right) (3.1992) \, e^{-n^{(1-\beta)}},$$

for all $j = 1, ..., m$.

At last we observe that

$$\left| G_n (f, x) - f (x) - \sum_{j=1}^{m} \left(\sum_{|\alpha|=j} \left(\frac{f_\alpha (x)}{\prod_{i=1}^{N} \alpha_i!} \right) G_n \left(\prod_{i=1}^{N} (\cdot - x_i)^{\alpha_i} , x \right) \right) \right| \le$$

$$(5.250312578)^N \cdot \left| G_n^* (f, x) - f (x) \sum_{k=\lceil na \rceil}^{\lfloor nb \rfloor} \Phi (nx - k) - \right.$$

$$\left. \sum_{j=1}^{m} \left(\sum_{|\alpha|=j} \left(\frac{f_\alpha (x)}{\prod_{i=1}^{N} \alpha_i!} \right) G_n^* \left(\prod_{i=1}^{N} (\cdot - x_i)^{\alpha_i} , x \right) \right) \right|.$$

Putting all of the above together we prove theorem. ∎

3.4 Complex Multivariate Neural Network Quantitative Approximations

We make

Remark 3.4. Let $X = \prod_{i=1}^{n} [a_i, b_i]$ or $\mathbb{R}^N$, and $f : X \to \mathbb{C}$ with real and imaginary parts $f_1, f_2 : f = f_1 + i f_2$, $i = \sqrt{-1}$. Clearly f is continuous iff f_1 and f_2 are continuous.

Given that $f_1, f_2 \in C^m (X)$, $m \in \mathbb{N}$, it holds

$$f_\alpha (x) = f_{1,\alpha} (x) + i f_{2,\alpha} (x), \tag{3.20}$$

where α denotes a partial derivative of any order and arrangement.

We denote by $C_B (\mathbb{R}^N, \mathbb{C})$ the space of continuous and bounded functions $f : \mathbb{R}^N \to \mathbb{C}$. Clearly f is bounded, iff both f_1, f_2 are bounded from $\mathbb{R}^N$ into $\mathbb{R}$, where $f = f_1 + i f_2$.

Here we define

$$G_n(f,x) := G_n(f_1,x) + iG_n(f_2,x), \quad x \in \left(\prod_{i=1}^{n}[a_i,b_i]\right), \qquad (3.21)$$

and

$$\overline{G}_n(f,x) := \overline{G}_n(f_1,x) + i\overline{G}_n(f_2,x), \quad x \in \mathbb{R}^N. \qquad (3.22)$$

We see here that

$$|G_n(f,x) - f(x)| \le |G_n(f_1,x) - f_1(x)| + |G_n(f_2,x) - f_2(x)|, \qquad (3.23)$$

and

$$\|G_n(f) - f\|_\infty \le \|G_n(f_1) - f_1\|_\infty + \|G_n(f_2) - f_2\|_\infty. \qquad (3.24)$$

Similarly we obtain

$$\left|\overline{G}_n(f,x) - f(x)\right| \le \left|\overline{G}_n(f_1,x) - f_1(x)\right| + \left|\overline{G}_n(f_2,x) - f_2(x)\right|, \quad x \in \mathbb{R}^N, \qquad (3.25)$$

and

$$\left\|\overline{G}_n(f) - f\right\|_\infty \le \left\|\overline{G}_n(f_1) - f_1\right\|_\infty + \left\|\overline{G}_n(f_2) - f_2\right\|_\infty. \qquad (3.26)$$

We give

Theorem 3.5. *Let* $f \in C\left(\prod_{i=1}^{n}[a_i,b_i],\mathbb{C}\right)$, $f = f_1 + if_2$, $0 < \beta < 1$, $n, N \in \mathbb{N}$, $x \in \left(\prod_{i=1}^{n}[a_i,b_i]\right)$. *Then*
i)
$$|G_n(f,x) - f(x)| \le (5.250312578)^N \cdot \qquad (3.27)$$

$$\left\{\omega_1\left(f_1,\frac{1}{n^\beta}\right) + \omega_1\left(f_2,\frac{1}{n^\beta}\right) + (6.3984)(\|f_1\|_\infty + \|f_2\|_\infty)e^{-n^{(1-\beta)}}\right\} =: \psi_1,$$

ii)
$$\|G_n(f) - f\|_\infty \le \psi_1. \qquad (3.28)$$

Proof. Use of Theorem 3.1 and Remark 3.4. ∎

We present

Theorem 3.6. *Let* $f \in C_B(\mathbb{R}^N,\mathbb{C})$, $f = f_1 + if_2$, $0 < \beta < 1$, $n, N \in \mathbb{N}$, $x \in \mathbb{R}^N$. *Then*
i)
$$\left|\overline{G}_n(f,x) - f(x)\right| \le \omega_1\left(f_1,\frac{1}{n^\beta}\right) + \omega_1\left(f_2,\frac{1}{n^\beta}\right) + \qquad (3.29)$$

$$(6.3984)(\|f_1\|_\infty + \|f_2\|_\infty)e^{-n^{(1-\beta)}} =: \psi_2,$$

ii)
$$\left\|\overline{G}_n(f) - f\right\|_\infty \le \psi_2. \qquad (3.30)$$

Proof. By Theorem 3.2 and Remark 3.4.   ∎

In the next we discuss high order of complex approximation by using the smoothness of f.

We present

Theorem 3.7. *Let* $f : \prod_{i=1}^{n} [a_i, b_i] \to \mathbb{C}$, *such that* $f = f_1 + if_2$. *Assume* $f_1, f_2 \in C^m \left(\prod_{i=1}^{n} [a_i, b_i] \right)$, $0 < \beta < 1$, $n, m, N \in \mathbb{N}$, $x \in \left(\prod_{i=1}^{n} [a_i, b_i] \right)$. *Then*
i)

$$\left| G_n\left(f, x\right) - f\left(x\right) - \sum_{j=1}^{m} \left(\sum_{|\alpha|=j} \left(\frac{f_\alpha\left(x\right)}{\prod_{i=1}^{N} \alpha_i!} \right) G_n\left(\prod_{i=1}^{N} \left(\cdot - x_i \right)^{\alpha_i}, x \right) \right) \right| \leq$$

$$(3.31)$$

$$(5.250312578)^N \cdot \left\{ \frac{N^m}{m! n^{m\beta}} \left(\omega_{1,m}^{\max}\left(f_{1,\alpha}, \frac{1}{n^\beta} \right) + \omega_{1,m}^{\max}\left(f_{2,\alpha}, \frac{1}{n^\beta} \right) \right) + \right.$$

$$\left. \left(\frac{(6.3984) \|b - a\|_\infty^m \left(\|f_{1,\alpha}\|_{\infty,m}^{\max} + \|f_{2,\alpha}\|_{\infty,m}^{\max} \right) N^m}{m!} \right) e^{-n^{(1-\beta)}} \right\},$$

ii)

$$|G_n\left(f, x\right) - f\left(x\right)| \leq (5.250312578)^N \cdot \qquad (3.32)$$

$$\left\{ \sum_{j=1}^{m} \left(\sum_{|\alpha|=j} \left(\frac{|f_{1,\alpha}\left(x\right)| + |f_{2,\alpha}\left(x\right)|}{\prod_{i=1}^{N} \alpha_i!} \right) \left[\frac{1}{n^{\beta j}} + \right. \right. \right.$$

$$\left. \left. \left(\prod_{i=1}^{N} \left(b_i - a_i\right)^{\alpha_i} \right) \cdot (3.1992)\, e^{-n^{(1-\beta)}} \right] \right) +$$

$$\frac{N^m}{m! n^{m\beta}} \left(\omega_{1,m}^{\max}\left(f_{1,\alpha}, \frac{1}{n^\beta} \right) + \omega_{1,m}^{\max}\left(f_{2,\alpha}, \frac{1}{n^\beta} \right) \right) +$$

$$\left(\frac{(6.3984) \|b - a\|_\infty^m \left(\|f_{1,\alpha}\|_{\infty,m}^{\max} + \|f_{2,\alpha}\|_{\infty,m}^{\max} \right) N^m}{m!} \right) e^{-n^{(1-\beta)}} \right\},$$

iii)

$$\|G_n\left(f\right) - f\|_\infty \leq (5.250312578)^N \cdot \qquad (3.33)$$

$$\left\{ \sum_{j=1}^{m} \left(\sum_{|\alpha|=j} \left(\frac{\|f_{1,\alpha}\left(x\right)\|_\infty + \|f_{2,\alpha}\left(x\right)\|_\infty}{\prod_{i=1}^{N} \alpha_i!} \right) \left[\frac{1}{n^{\beta j}} + \right. \right. \right.$$

$$\left. \left. \left(\prod_{i=1}^{N} \left(b_i - a_i\right)^{\alpha_i} \right) \cdot (3.1992)\, e^{-n^{(1-\beta)}} \right] \right) +$$

$$\frac{N^m}{m!n^{m\beta}} \left(\omega_{1,m}^{\max} \left(f_{1,\alpha}, \frac{1}{n^\beta} \right) + \omega_{1,m}^{\max} \left(f_{2,\alpha}, \frac{1}{n^\beta} \right) \right) +$$

$$+ \left(\frac{(6.3984) \|b - a\|_\infty^m \left(\|f_{1,\alpha}\|_{\infty,m}^{\max} + \|f_{2,\alpha}\|_{\infty,m}^{\max} \right) N^m}{m!} \right) e^{-n^{(1-\beta)}} \Bigg\},$$

iv) Suppose $f_\alpha(x_0) = 0$, *for all* $\alpha : |\alpha| = 1, ..., m$; $x_0 \in \left(\prod_{i=1}^N [a_i, b_i] \right)$.
Then

$$|G_n(f, x_0) - f(x_0)| \le (5.250312578)^N \cdot \tag{3.34}$$

$$\left\{ \frac{N^m}{m!n^{m\beta}} \left(\omega_{1,m}^{\max} \left(f_{1,\alpha}, \frac{1}{n^\beta} \right) + \omega_{1,m}^{\max} \left(f_{2,\alpha}, \frac{1}{n^\beta} \right) \right) + \right.$$

$$\left(\frac{(6.3984) \|b - a\|_\infty^m \left(\|f_{1,\alpha}\|_{\infty,m}^{\max} + \|f_{2,\alpha}\|_{\infty,m}^{\max} \right) N^m}{m!} \right) e^{-n^{(1-\beta)}} \Bigg\},$$

notice in the last the extremely high rate of convergence at $n^{-\beta(m+1)}$.

Proof. By Theorem 3.3 and Remark 3.4.   ∎

Example 3.8.
Consider $f(x, y) = e^{x+y}$, $(x, y) \in [-1, 1]^2$. Let $\overline{x} = (x_1, y_1)$, $\overline{y} = (x_2, y_2)$, we see that

$$
\begin{aligned}
|f(\overline{x}) - f(\overline{y})| &= \left| e^{x_1+y_1} - e^{x_2+y_2} \right| \\
&= \left| e^{x_1} e^{y_1} - e^{x_1} e^{y_2} + e^{x_1} e^{y_2} - e^{x_2} e^{y_2} \right| \\
&= \left| e^{x_1} (e^{y_1} - e^{y_2}) + e^{y_2} (e^{x_1} - e^{x_2}) \right| \\
&\le e \left(|e^{y_1} - e^{y_2}| + |e^{x_1} - e^{x_2}| \right) \\
&\le e^2 \left[|y_1 - y_2| + |x_1 - x_2| \right] \\
&\le 2e^2 \|\overline{x} - \overline{y}\|_\infty.
\end{aligned}
$$

That is

$$|f(\overline{x}) - f(\overline{y})| \le 2e^2 \|\overline{x} - \overline{y}\|_\infty. \tag{3.35}$$

Consequently by (3.7) we get that

$$\omega_1(f, h) \le 2e^2 h, \quad h > 0. \tag{3.36}$$

Therefore by (3.13) we derive

$$\left\| G_n \left(e^{x+y} \right) (x, y) - e^{x+y} \right\|_\infty \le (27.5657821) \tag{3.37}$$

$$\cdot e^2 \left\{ \frac{2}{n^\beta} + (6.3984) e^{-n^{(1-\beta)}} \right\},$$

where $0 < \beta < 1$ and $n \in \mathbb{N}$.

Example 3.9

Let $f(x_1, \ldots, x_N) = \sum_{i=1}^{N} \sin x_i$, $(x_1, \ldots, x_N) \in \mathbb{R}^N$, $N \in \mathbb{N}$. Denote $\overline{x} = (x_1, \ldots, x_N)$, $\overline{y} = (y_1, \ldots, y_N)$ and see that

$$\left| \sum_{i=1}^{N} \sin x_i - \sum_{i=1}^{N} \sin y_i \right| \leq \sum_{i=1}^{N} |\sin x_i - \sin y_i|$$

$$\leq \sum_{i=1}^{N} |x_i - y_i|$$

$$\leq N \|\overline{x} - \overline{y}\|_{\infty}.$$

That is

$$|f(\overline{x}) - f(\overline{y})| \leq N \|\overline{x} - \overline{y}\|_{\infty}. \tag{3.38}$$

Consequently by (3.7) we obtain that

$$\omega_1(f, h) \leq Nh, \ h > 0. \tag{3.39}$$

Therefore by (3.15) we derive

$$\left\| \overline{G_n} \left(\sum_{i=1}^{N} \sin x_i \right) (x_1, \ldots, x_N) - \sum_{i=1}^{N} \sin x_i \right\|_{\infty} \leq N \left(\frac{1}{n^{\beta}} + (6.3984) e^{-n^{(1-\beta)}} \right),$$

$$\tag{3.40}$$

where $0 < \beta < 1$ and $n \in \mathbb{N}$.

One can easily construct many other interesting examples.

References

[1] Anastassiou, G.A.: Rate of convergence of some neural network operators to the unit-univariate case. J. Math. Anal. Appli. 212, 237–262 (1997)

[2] Anastassiou, G.A.: Rate of convergence of some multivariate neural network operators to the unit. J. Comp. Math. Appl. 40, 1–19 (2000)

[3] Anastassiou, G.A.: Quantitative Approximations. Chapman&Hall/CRC, Boca Raton (2001)

[4] Anastassiou, G.A.: Univariate sigmoidal neural network approximation. Journal of Comp. Anal. and Appl., (accepted 2011)

[5] Anastassiou, G.A.: Multivariate sigmoidal neural network approximation. Neural Networks 24, 378–386 (2011)

[6] Barron, A.R.: Universal approximation bounds for superpositions of a sigmoidal function. IEEE Trans. Inform. Theory 39, 930–945 (1993)

[7] Brauer, F., Castillo-Chavez, C.: Mathematical models in population biology and epidemiology, pp. 8–9. Springer, New York (2001)

[8] Cao, F.L., Xie, T.F., Xu, Z.B.: The estimate for approximation error of neural networks: a constructive approach. Neurocomputing 71, 626–630 (2008)

[9] Chen, T.P., Chen, H.: Universal approximation to nonlinear operators by neural networks with arbitrary activation functions and its applications to a dynamic system. IEEE Trans. Neural Networks 6, 911–917 (1995)

[10] Chen, Z., Cao, F.: The approximation operators with sigmoidal functions. Computers and Mathematics with Applications 58, 758–765 (2009)

[11] Chui, C.K., Li, X.: Approximation by ridge functions and neural networks with one hidden layer. J. Approx. Theory 70, 131–141 (1992)

[12] Cybenko, G.: Approximation by superpositions of sigmoidal function. Math. of Control Signals and System 2, 303–314 (1989)

[13] Ferrari, S., Stengel, R.F.: Smooth function approximation using neural networks. IEEE Trans. Neural Networks 16, 24–38 (2005)

[14] Funahashi, K.I.: On the approximate realization of continuous mappings by neural networks. Neural Networks 2, 183–192 (1989)

[15] Hahm, N., Hong, B.I.: An approximation by neural networks with a fixed weight. Computers & Math. with Appli. 47, 1897–1903 (2004)

[16] Hornik, K., Stinchombe, M., White, H.: Multilayer feedforward networks are universal approximators. Neural Networks 2, 359–366 (1989)

[17] Hornik, K., Stinchombe, M., White, H.: Universal approximation of an un-known mapping and its derivatives using multilayer feedforward networks. Neural Networks 3, 551–560 (1990)

[18] Hritonenko, N., Yatsenko, Y.: Mathematical modeling in economics, ecology and the environment, pp. 92–93. Science Press, Beijing (2006) (reprint)

[19] Leshno, M., Lin, V.Y., Pinks, A., Schocken, S.: Multilayer feedforward net-works with a nonpolynomial activation function can approximate any function. Neural Networks 6, 861–867 (1993)

[20] Lorentz, G.G.: Approximation of Functions. Rinehart and Winston, New York (1966)

[21] Maiorov, V., Meir, R.S.: Approximation bounds for smooth functions in $C(R^d)$ by neural and mixture networks. IEEE Trans. Neural Networks 9, 969–978 (1998)

[22] Makovoz, Y.: Uniform approximation by neural networks. J. Approx. The-ory 95, 215–228 (1998)

[23] Mhaskar, H.N., Micchelli, C.A.: Approximation by superposition of a sigmoidal function. Adv. Applied Math. 13, 350–373 (1992)

[24] Mhaskar, H.N., Micchelli, C.A.: Degree of approximation by neural networks with a single hidden layer. Adv. Applied Math. 16, 151–183 (1995)

[25] Suzuki, S.: Constructive function approximation by three-layer artificial neural networks. Neural Networks 11, 1049–1058 (1998)

[26] Xie, T.F., Zhou, S.P.: Approximation Theory of Real Functions. Hangzhou University Press, Hangzhou (1998)

[27] Xu, Z.B., Cao, F.L.: The essential order of approximation for neural networks. Science in China (Ser. F) 47, 97–112 (2004)

Chapter 4
Multivariate Hyperbolic Tangent Neural Network Quantitative Approximation

Here we give the multivariate quantitative approximation of real and complex valued continuous multivariate functions on a box or $\mathbb{R}^N$, $N \in \mathbb{N}$, by the multivariate quasi-interpolation hyperbolic tangent neural network operators. This approximation is obtained by establishing multidimensional Jackson type inequalities involving the multivariate modulus of continuity of the engaged function or its high order partial derivatives. The multivariate operators are defined by using a multidimensional density function induced by the hyperbolic tangent function. Our approximations are pointwise and uniform. The related feed-forward neural network is with one hidden layer. This chapter is based on [6].

4.1 Introduction

The author in [1], [2], and [3], see chapters 2-5, was the first to establish neural network approximations to continuous functions with rates by very specifically defined neural network operators of Cardaliagnet-Euvrard and "Squashing" types, by employing the modulus of continuity of the engaged function or its high order derivative, and producing very tight Jackson type inequalities. He treats there both the univariate and multivariate cases. The defining these operators "bell-shaped" and "squashing" functions are assumed to be of compact support. Also in [3] he gives the Nth order asymptotic expansion for the error of weak approximation of these two operators to a special natural class of smooth functions, see chapters 4-5 there.

For this chapter the author is inspired by the article [7] by Z. Chen and F. Cao, also by [4], [5].

G.A. Anastassiou: Intelligent Systems: Approximation by ANN, ISRL 19, pp. 89–107.
springerlink.com   © Springer-Verlag Berlin Heidelberg 2011

The author here performs multivariate hyperbolic tangent neural network approximations to continuous functions over boxes or over the whole $\mathbb{R}^N$, $N \in \mathbb{N}$, then he extends his results to complex valued multivariate functions. All convergences here are with rates expressed via the multivariate modulus of continuity of the involved function or its high order partial derivative, and given by very tight multidimensional Jackson type inequalities.

The author here comes up with the "right" precisely defined multivariate quasi-interpolation neural network operators related to boxes or $\mathbb{R}^N$. The boxes are not necessarily symmetric to the origin. In preparation to prove our results we prove important properties of the basic multivariate density function induced by hyperbolic tangent function and defining our operators.

Feed-forward neural networks (FNNs) with one hidden layer, the only type of networks we deal with in this chapter, are mathematically expressed as

$$N_n(x) = \sum_{j=0}^{n} c_j \sigma \left(\langle a_j \cdot x \rangle + b_j \right), \quad x \in \mathbb{R}^s, \quad s \in \mathbb{N},$$

where for $0 \leq j \leq n$, $b_j \in \mathbb{R}$ are the thresholds, $a_j \in \mathbb{R}^s$ are the connection weights, $c_j \in \mathbb{R}$ are the coefficients, $\langle a_j \cdot x \rangle$ is the inner product of a_j and x, and σ is the activation function of the network. In many fundamental network models, the activation function is the hyperbolic tangent. About neural networks see [8], [9], [10].

4.2 Basic Ideas

We consider here the hyperbolic tangent function $\tanh x$, $x \in \mathbb{R}$:

$$\tanh x := \frac{e^x - e^{-x}}{e^x + e^{-x}}.$$

It has the properties $\tanh 0 = 0$, $-1 < \tanh x < 1$, $\forall\, x \in \mathbb{R}$, and $\tanh(-x) = -\tanh x$. Furthermore $\tanh x \to 1$ as $x \to \infty$, and $\tanh x \to -1$, as $x \to -\infty$, and it is strictly increasing on $\mathbb{R}$.

This function plays the role of an activation function in the hidden layer of neural networks.

We further consider

$$\Psi(x) := \frac{1}{4} \left(\tanh(x + 1) - \tanh(x - 1) \right) > 0, \quad \forall\, x \in \mathbb{R}.$$

We easily see that $\Psi(-x) = \Psi(x)$, that is Ψ is even on $\mathbb{R}$. Obviously Ψ is differentiable, thus continuous.

Proposition 4.1. *([5]) $\Psi(x)$ for $x \geq 0$ is strictly decreasing.*

Obviously $\Psi(x)$ is strictly increasing for $x \leq 0$. Also it holds $\lim\limits_{x \to -\infty} \Psi(x) = 0 = \lim\limits_{x \to \infty} \Psi(x)$.

Infact Ψ has the bell shape with horizontal asymptote the x-axis. So the maximum of Ψ is zero, $\Psi(0) = 0.3809297$.

Theorem 4.2. *([5]) We have that $\sum_{i=-\infty}^{\infty} \Psi(x - i) = 1$, $\forall x \in \mathbb{R}$.*

Therefore

$$\sum_{i=-\infty}^{\infty} \Psi(nx - i) = 1, \quad \forall n \in \mathbb{N}, \forall x \in \mathbb{R}.$$

Also it holds

$$\sum_{i=-\infty}^{\infty} \Psi(x + i) = 1, \quad \forall x \in \mathbb{R}.$$

Theorem 4.3. *([5]) It holds $\int_{-\infty}^{\infty} \Psi(x)\, dx = 1$.*

So $\Psi(x)$ is a density function on $\mathbb{R}$.

Theorem 4.4. *([5]) Let $0 < \alpha < 1$ and $n \in \mathbb{N}$. It holds*

$$\sum_{\substack{k = -\infty \\ : |nx - k| \geq n^{1-\alpha}}}^{\infty} \Psi(nx - k) \leq e^4 \cdot e^{-2n^{(1-\alpha)}}.$$

Denote by $\lfloor \cdot \rfloor$ the integral part of the number and by $\lceil \cdot \rceil$ the ceiling of the number.

Theorem 4.5. *([5]) Let $x \in [a, b] \subset \mathbb{R}$ and $n \in \mathbb{N}$ so that $\lceil na \rceil \leq \lfloor nb \rfloor$. It holds*

$$\frac{1}{\sum_{k=\lceil na \rceil}^{\lfloor nb \rfloor} \Psi(nx - k)} < \frac{1}{\Psi(1)} = 4.1488766.$$

Also by [5] we get that

$$\lim_{n \to \infty} \sum_{k=\lceil na \rceil}^{\lfloor nb \rfloor} \Psi(nx - k) \neq 1,$$

for at least some $x \in [a, b]$.

In this chapter we employ

$$\Theta(x_1, ..., x_N) := \Theta(x) := \prod_{i=1}^{N} \Psi(x_i), \quad x = (x_1, ..., x_N) \in \mathbb{R}^N, N \in \mathbb{N}. \quad (4.1)$$

It has the properties:

(i) $\Theta(x) > 0, \ \forall \ x \in \mathbb{R}^N$,

(ii)

$$\sum_{k=-\infty}^{\infty} \Theta(x-k) := \sum_{k_1=-\infty}^{\infty} \sum_{k_2=-\infty}^{\infty} \dots \sum_{k_N=-\infty}^{\infty} \Theta(x_1-k_1, \dots, x_N-k_N) = 1,$$

where $k := (k_1, \dots, k_n), \ \forall \ x \in \mathbb{R}^N$.

(iii)

$$\sum_{k=-\infty}^{\infty} \Theta(nx-k) :=$$

$$\sum_{k_1=-\infty}^{\infty} \sum_{k_2=-\infty}^{\infty} \dots \sum_{k_N=-\infty}^{\infty} \Theta(nx_1-k_1, \dots, nx_N-k_N) = 1,$$

$\forall \ x \in \mathbb{R}^N; \ n \in \mathbb{N}$.

(iv)

$$\int_{\mathbb{R}^N} \Theta(x) \, dx = 1,$$

that is Θ is a multivariate density function.

Here $\|x\|_{\infty} := \max\{|x_1|, \dots, |x_N|\}, \ x \in \mathbb{R}^N$, also set $\infty := (\infty, \dots, \infty)$, $-\infty := (-\infty, \dots, -\infty)$ upon the multivariate context, and

$$\lceil na \rceil := (\lceil na_1 \rceil, \dots, \lceil na_N \rceil),$$
$$\lfloor nb \rfloor := (\lfloor nb_1 \rfloor, \dots, \lfloor nb_N \rfloor),$$

where $a := (a_1, \dots, a_N), \ b := (b_1, \dots, b_N)$.

We clearly see that

$$\sum_{k=\lceil na \rceil}^{\lfloor nb \rfloor} \Theta(nx-k) = \sum_{k=\lceil na \rceil}^{\lfloor nb \rfloor} \prod_{i=1}^{N} \Psi(nx_i - x_i) =$$

$$\sum_{k_1=\lceil na_1 \rceil}^{\lfloor nb_1 \rfloor} \dots \sum_{k_N=\lceil na_N \rceil}^{\lfloor nb_N \rfloor} \prod_{i=1}^{N} \Psi(nx_i - k_i) = \prod_{i=1}^{N} \left(\sum_{k_i=\lceil na_i \rceil}^{\lfloor nb_i \rfloor} \Psi(nx_i - k_i) \right).$$

For $0 < \beta < 1$ and $n \in \mathbb{N}$, fixed $x \in \mathbb{R}^N$, we have that

$$\sum_{k=\lceil na \rceil}^{\lfloor nb \rfloor} \Theta(nx-k) =$$

$$\sum_{\substack{k = \lceil na \rceil \\ \left\| \frac{k}{n} - x \right\|_\infty \le \frac{1}{n^\beta}}}^{\lfloor nb \rfloor} \Theta\left(nx - k\right) + \sum_{\substack{k = \lceil na \rceil \\ \left\| \frac{k}{n} - x \right\|_\infty > \frac{1}{n^\beta}}}^{\lfloor nb \rfloor} \Theta\left(nx - k\right).$$

In the last two sums the counting is over disjoint vector sets of k's, because the condition $\left\| \frac{k}{n} - x \right\|_\infty > \frac{1}{n^\beta}$ implies that there exists at least one $\left| \frac{k_r}{n} - x_r \right| > \frac{1}{n^\beta}$, $r \in \{1, ..., N\}$.

We treat

$$\sum_{\substack{k = \lceil na \rceil \\ \left\| \frac{k}{n} - x \right\|_\infty > \frac{1}{n^\beta}}}^{\lfloor nb \rfloor} \Theta\left(nx - k\right) = \prod_{i=1}^{N} \left(\sum_{\substack{k_i = \lceil na_i \rceil \\ \left\| \frac{k}{n} - x \right\|_\infty > \frac{1}{n^\beta}}}^{\lfloor nb_i \rfloor} \Psi\left(nx_i - k_i\right) \right)$$

$$\le \left(\prod_{\substack{i=1 \\ i \ne r}}^{N} \left(\sum_{k_i = -\infty}^{\infty} \Psi\left(nx_i - k_i\right) \right) \right) \cdot \left(\sum_{\substack{k_r = \lceil na_r \rceil \\ \left| \frac{k_r}{n} - x_r \right| > \frac{1}{n^\beta}}}^{\lfloor nb_r \rfloor} \Psi\left(nx_r - k_r\right) \right)$$

$$= \left(\sum_{\substack{k_r = \lceil na_r \rceil \\ \left| \frac{k_r}{n} - x_r \right| > \frac{1}{n^\beta}}}^{\lfloor nb_r \rfloor} \Psi\left(nx_r - k_r\right) \right)$$

$$\le \sum_{\substack{k_r = -\infty \\ \left| \frac{k_r}{n} - x_r \right| > \frac{1}{n^\beta}}}^{\infty} \Psi\left(nx_r - k_r\right) \overset{\text{(by Theorem 4.4)}}{\le} e^4 \cdot e^{-2n^{(1-\beta)}}.$$

We have established that

(v)

$$\sum_{\substack{k = \lceil na \rceil \\ \left\| \frac{k}{n} - x \right\|_\infty > \frac{1}{n^\beta}}}^{\lfloor nb \rfloor} \Theta\left(nx - k\right) \le e^4 \cdot e^{-2n^{(1-\beta)}},$$

$0 < \beta < 1$, $n \in \mathbb{N}$, $x \in \left(\prod_{i=1}^{N} [a_i, b_i] \right)$.

By Theorem 4.5 clearly we obtain

$$0 < \frac{1}{\sum_{k=\lceil na \rceil}^{\lfloor nb \rfloor} \Theta (nx - k)} = \frac{1}{\prod_{i=1}^{N} \left(\sum_{k_i=\lceil na_i \rceil}^{\lfloor nb_i \rfloor} \Psi (nx_i - k_i) \right)}$$

$$< \frac{1}{(\Psi (1))^N} = (4.1488766)^N .$$

That is,

(vi) it holds

$$0 < \frac{1}{\sum_{k=\lceil na \rceil}^{\lfloor nb \rfloor} \Theta (nx - k)} < \frac{1}{(\Psi (1))^N} = (4.1488766)^N ,$$

$$\forall \, x \in \left(\prod_{i=1}^{N} [a_i, b_i] \right), \ n \in \mathbb{N}.$$

It is also obvious that

(vii)

$$\sum_{\substack{k = -\infty \\ \left\{ \left\| \frac{k}{n} - x \right\|_\infty > \frac{1}{n^\beta} \right.}}^{\infty} \Theta (nx - k) \le e^4 \cdot e^{-2n^{(1-\beta)}},$$

$$0 < \beta < 1, \, n \in \mathbb{N}, \, x \in \mathbb{R}^N.$$

Also we find

$$\lim_{n \to \infty} \sum_{k=\lceil na \rceil}^{\lfloor nb \rfloor} \Theta (nx - k) \ne 1,$$

for at least some $x \in \left(\prod_{i=1}^{N} [a_i, b_i] \right)$.

Let $f \in C \left(\prod_{i=1}^{N} [a_i, b_i] \right)$ and $n \in \mathbb{N}$ such that $\lceil na_i \rceil \le \lfloor nb_i \rfloor$, $i = 1, ..., N$. We introduce and define the multivariate positive linear neural network operator $(x := (x_1, ..., x_N) \in \left(\prod_{i=1}^{N} [a_i, b_i] \right))$

$$F_n (f, x_1, ..., x_N) := F_n (f, x) := \frac{\sum_{k=\lceil na \rceil}^{\lfloor nb \rfloor} f \left(\frac{k}{n} \right) \Theta (nx - k)}{\sum_{k=\lceil na \rceil}^{\lfloor nb \rfloor} \Theta (nx - k)} \qquad (4.2)$$

$$:= \frac{\sum_{k_1=\lceil na_1 \rceil}^{\lfloor nb_1 \rfloor} \sum_{k_2=\lceil na_2 \rceil}^{\lfloor nb_2 \rfloor} \cdots \sum_{k_N=\lceil na_N \rceil}^{\lfloor nb_N \rfloor} f \left(\frac{k_1}{n}, ..., \frac{k_N}{n} \right) \left(\prod_{i=1}^{N} \Psi (nx_i - k_i) \right)}{\prod_{i=1}^{N} \left(\sum_{k_i=\lceil na_i \rceil}^{\lfloor nb_i \rfloor} \Psi (nx_i - k_i) \right)}.$$

For large enough n we always obtain $\lceil na_i \rceil \le \lfloor nb_i \rfloor$, $i = 1, ..., N$. Also $a_i \le \frac{k_i}{n} \le b_i$, iff $\lceil na_i \rceil \le k_i \le \lfloor nb_i \rfloor$, $i = 1, ..., N$.

We study here the pointwise and uniform convergence of $F_n(f)$ to f with rates.

For convenience we call

$$F_n^*(f,x) := \sum_{k=\lceil na \rceil}^{\lfloor nb \rfloor} f\left(\frac{k}{n}\right) \Theta(nx-k) \tag{4.3}$$

$$:= \sum_{k_1=\lceil na_1 \rceil}^{\lfloor nb_1 \rfloor} \sum_{k_2=\lceil na_2 \rceil}^{\lfloor nb_2 \rfloor} \cdots \sum_{k_N=\lceil na_N \rceil}^{\lfloor nb_N \rfloor} f\left(\frac{k_1}{n}, ..., \frac{k_N}{n}\right) \left(\prod_{i=1}^{N} \Psi(nx_i - k_i)\right),$$

$\forall\, x \in \left(\prod_{i=1}^{N} [a_i, b_i]\right).$

That is

$$F_n(f,x) := \frac{F_n^*(f,x)}{\sum_{k=\lceil na \rceil}^{\lfloor nb \rfloor} \Theta(nx-k)}, \tag{4.4}$$

$\forall\, x \in \left(\prod_{i=1}^{N} [a_i, b_i]\right),\; n \in \mathbb{N}.$

So that

$$F_n(f,x) - f(x) = \frac{F_n^*(f,x) - f(x) \sum_{k=\lceil na \rceil}^{\lfloor nb \rfloor} \Theta(nx-k)}{\sum_{k=\lceil na \rceil}^{\lfloor nb \rfloor} \Theta(nx-k)}. \tag{4.5}$$

Consequently we obtain

$$|F_n(f,x) - f(x)| \leq (4.1488766)^N \left| F_n^*(f,x) - f(x) \sum_{k=\lceil na \rceil}^{\lfloor nb \rfloor} \Theta(nx-k) \right|, \tag{4.6}$$

$\forall\, x \in \left(\prod_{i=1}^{N} [a_i, b_i]\right).$

We will estimate the right hand side of (4.6).

For that we need, for $f \in C\left(\prod_{i=1}^{N} [a_i, b_i]\right)$ the first multivariate modulus of continuity

$$\omega_1(f,h) := \sup_{\substack{x,y \in \prod_{i=1}^{N} [a_i, b_i] \\ \|x-y\|_\infty \leq h}} |f(x) - f(y)|,\quad h > 0. \tag{4.7}$$

Similarly it is defined for $f \in C_B(\mathbb{R}^N)$ (continuous and bounded functions on $\mathbb{R}^N$). We have that $\lim_{h \to 0} \omega_1(f,h) = 0$.

When $f \in C_B(\mathbb{R}^N)$ we define,

$$\overline{F}_n(f,x) := \overline{F}_n(f,x_1, ..., x_N) := \sum_{k=-\infty}^{\infty} f\left(\frac{k}{n}\right) \Theta(nx-k) := \tag{4.8}$$

$$\sum_{k_1=-\infty}^{\infty} \sum_{k_2=-\infty}^{\infty} \cdots \sum_{k_N=-\infty}^{\infty} f\left(\frac{k_1}{n}, \frac{k_2}{n}, ..., \frac{k_N}{n}\right) \left(\prod_{i=1}^{N} \Psi\left(nx_i - k_i\right)\right),$$

$n \in \mathbb{N}, \forall x \in \mathbb{R}^N, N \geq 1$, the multivariate quasi-interpolation neural network operator.

Notice here that for large enough $n \in \mathbb{N}$ we get that

$$e^{-2n^{(1-\beta)}} < n^{-\beta j}, \quad j = 1, ..., m \in \mathbb{N}, \quad 0 < \beta < 1. \tag{4.9}$$

Thus be given fixed $A, B > 0$, for the linear combination $\left(An^{-\beta j} + Be^{-2n^{(1-\beta)}}\right)$ the (dominant) rate of convergence to zero is $n^{-\beta j}$. The closer β is to 1 we get faster and better rate of convergence to zero.

Let $f \in C^m\left(\prod_{i=1}^{N}[a_i, b_i]\right)$, $m, N \in \mathbb{N}$. Here f_α denotes a partial derivative of f, $\alpha := (\alpha_1, ..., \alpha_N)$, $\alpha_i \in \mathbb{Z}^+$, $i = 1, ..., N$, and $|\alpha| := \sum_{i=1}^{N} \alpha_i = l$, where $l = 0, 1, ..., m$. We write also $f_\alpha := \frac{\partial^\alpha f}{\partial x^\alpha}$ and we say it is of order l.

We denote

$$\omega_{1,m}^{\max}(f_\alpha, h) := \max_{\alpha:|\alpha|=m} \omega_1(f_\alpha, h). \tag{4.10}$$

Call also

$$\|f_\alpha\|_{\infty,m}^{\max} := \max_{|\alpha|=m} \{\|f_\alpha\|_\infty\}, \tag{4.11}$$

$\|\cdot\|_\infty$ is the supremum norm.

4.3   Real Multivariate Neural Network Quantitative Approximations

Here we show a series of multivariate neural network approximations to a function given with rates.

We first present

Theorem 4.6. *Let* $f \in C\left(\prod_{i=1}^{N}[a_i, b_i]\right)$, $0 < \beta < 1$, $x \in \left(\prod_{i=1}^{N}[a_i, b_i]\right)$, $n, N \in \mathbb{N}$. *Then*
  i)

$$|F_n(f, x) - f(x)| \leq (4.1488766)^N \cdot$$

$$\left\{\omega_1\left(f, \frac{1}{n^\beta}\right) + 2e^4 \|f\|_\infty e^{-2n^{(1-\beta)}}\right\} =: \lambda_1, \tag{4.12}$$

  ii)

$$\|F_n(f) - f\|_\infty \leq \lambda_1. \tag{4.13}$$

Proof. We see that

$$\Delta(x) := F_n^*(f, x) - f(x) \sum_{k=\lceil na \rceil}^{\lfloor nb \rfloor} \Theta(nx - k) =$$

$$\sum_{k=\lceil na \rceil}^{\lfloor nb \rfloor} f\left(\frac{k}{n}\right) \Theta(nx - k) - \sum_{k=\lceil na \rceil}^{\lfloor nb \rfloor} f(x) \Theta(nx - k) =$$

$$\sum_{k=\lceil na \rceil}^{\lfloor nb \rfloor} \left(f\left(\frac{k}{n}\right) - f(x) \right) \Theta(nx - k).$$

Therefore

$$|\Delta(x)| \le \sum_{k=\lceil na \rceil}^{\lfloor nb \rfloor} \left| f\left(\frac{k}{n}\right) - f(x) \right| \Theta(nx - k) =$$

$$\sum_{\substack{k=\lceil na \rceil \\ \left\| \frac{k}{n} - x \right\|_\infty \le \frac{1}{n^\beta}}}^{\lfloor nb \rfloor} \left| f\left(\frac{k}{n}\right) - f(x) \right| \Theta(nx - k) +$$

$$\sum_{\substack{k=\lceil na \rceil \\ \left\| \frac{k}{n} - x \right\|_\infty > \frac{1}{n^\beta}}}^{\lfloor nb \rfloor} \left| f\left(\frac{k}{n}\right) - f(x) \right| \Theta(nx - k) \le$$

$$\omega_1\left(f, \frac{1}{n^\beta}\right) + 2\|f\|_\infty \sum_{\substack{k=\lceil na \rceil \\ \left\| \frac{k}{n} - x \right\|_\infty > \frac{1}{n^\beta}}}^{\lfloor nb \rfloor} \Theta(nx - k) \le$$

$$\omega_1\left(f, \frac{1}{n^\beta}\right) + 2e^4 \|f\|_\infty e^{-2n^{(1-\beta)}}.$$

So that

$$|\Delta(x)| \le \omega_1\left(f, \frac{1}{n^\beta}\right) + 2e^4 \|f\|_\infty e^{-2n^{(1-\beta)}}.$$

Now using (4.6) we prove claim. ∎

Next we give

Theorem 4.7. Let $f \in C_B(\mathbb{R}^N)$, $0 < \beta < 1$, $x \in \mathbb{R}^N$, $n, N \in \mathbb{N}$. Then
i)

$$\left| \overline{F}_n(f, x) - f(x) \right| \le \omega_1\left(f, \frac{1}{n^\beta}\right) + 2e^4 \|f\|_\infty e^{-2n^{(1-\beta)}} =: \lambda_2, \qquad (4.14)$$

ii)
$$\left\| \overline{F}_n \left(f \right) - f \right\|_\infty \le \lambda_2. \tag{4.15}$$

Proof. We have

$$\overline{F}_n \left(f, x \right) := \sum_{k=-\infty}^{\infty} f \left(\frac{k}{n} \right) \Theta \left(nx - k \right).$$

Therefore

$$E_n \left(x \right) := \overline{F}_n \left(f, x \right) - f \left(x \right) = \sum_{k=-\infty}^{\infty} f \left(\frac{k}{n} \right) \Theta \left(nx - k \right) - f \left(x \right) \sum_{k=-\infty}^{\infty} \Theta \left(nx - k \right) =$$

$$\sum_{k=-\infty}^{\infty} \left(f \left(\frac{k}{n} \right) - f \left(x \right) \right) \Theta \left(nx - k \right).$$

Hence

$$\left| E_n \left(x \right) \right| \le \sum_{k=-\infty}^{\infty} \left| f \left(\frac{k}{n} \right) - f \left(x \right) \right| \Theta \left(nx - k \right) =$$

$$\sum_{\substack{k=-\infty \\ \left\| \frac{k}{n} - x \right\|_\infty \le \frac{1}{n^\beta}}}^{\infty} \left| f \left(\frac{k}{n} \right) - f \left(x \right) \right| \Theta \left(nx - k \right) +$$

$$\sum_{\substack{k=-\infty \\ \left\| \frac{k}{n} - x \right\|_\infty > \frac{1}{n^\beta}}}^{\infty} \left| f \left(\frac{k}{n} \right) - f \left(x \right) \right| \Theta \left(nx - k \right) \le$$

$$\omega_1 \left(f, \frac{1}{n^\beta} \right) + 2 \left\| f \right\|_\infty \sum_{\substack{k=-\infty \\ \left\| \frac{k}{n} - x \right\|_\infty > \frac{1}{n^\beta}}}^{\infty} \Theta \left(nx - k \right) \le$$

$$\omega_1 \left(f, \frac{1}{n^\beta} \right) + 2e^4 \left\| f \right\|_\infty e^{-2n^{(1-\beta)}}.$$

Thus

$$\left| E_n \left(x \right) \right| \le \omega_1 \left(f, \frac{1}{n^\beta} \right) + 2e^4 \left\| f \right\|_\infty e^{-2n^{(1-\beta)}},$$

proving the claim.   ∎

In the next we discuss high order of approximation by using the smoothness of f.

We present

Theorem 4.8. *Let* $f \in C^m \left(\prod_{i=1}^{N} [a_i, b_i] \right)$, $0 < \beta < 1$, $n, m, N \in \mathbb{N}$, $x \in$ $\left(\prod_{i=1}^{N} [a_i, b_i] \right)$. *Then*

i)

$$
\left| F_n \left(f, x \right) - f \left(x \right) - \sum_{j=1}^{m} \left(\sum_{|\alpha|=j} \left(\frac{f_\alpha \left(x \right)}{\prod_{i=1}^{N} \alpha_i!} \right) F_n \left(\prod_{i=1}^{N} \left(\cdot - x_i \right)^{\alpha_i}, x \right) \right) \right| \leq
$$
(4.16)

$$
(4.1488766)^N \cdot \left\{ \frac{N^m}{m! n^{m\beta}} \omega_{1,m}^{\max} \left(f_\alpha, \frac{1}{n^\beta} \right) + \left(\frac{2e^4 \|b - a\|_\infty^m \|f_\alpha\|_{\infty,m}^{\max} N^m}{m!} \right) e^{-2n^{(1-\beta)}} \right\},
$$

ii)

$$
|F_n \left(f, x \right) - f \left(x \right)| \leq (4.1488766)^N \cdot
$$
(4.17)

$$
\left\{ \sum_{j=1}^{m} \left(\sum_{|\alpha|=j} \left(\frac{|f_\alpha \left(x \right)|}{\prod_{i=1}^{N} \alpha_i!} \right) \left[\frac{1}{n^{\beta j}} + \left(\prod_{i=1}^{N} \left(b_i - a_i \right)^{\alpha_i} \right) \cdot e^4 e^{-2n^{(1-\beta)}} \right] \right) + \frac{N^m}{m! n^{m\beta}} \omega_{1,m}^{\max} \left(f_\alpha, \frac{1}{n^\beta} \right) + \left(\frac{2e^4 \|b - a\|_\infty^m \|f_\alpha\|_{\infty,m}^{\max} N^m}{m!} \right) e^{-2n^{(1-\beta)}} \right\},
$$

iii)

$$
\|F_n \left(f \right) - f\|_\infty \leq (4.1488766)^N \cdot
$$
(4.18)

$$
\left\{ \sum_{j=1}^{m} \left(\sum_{|\alpha|=j} \left(\frac{\|f_\alpha\|_\infty}{\prod_{i=1}^{N} \alpha_i!} \right) \left[\frac{1}{n^{\beta j}} + \left(\prod_{i=1}^{N} \left(b_i - a_i \right)^{\alpha_i} \right) e^4 e^{-2n^{(1-\beta)}} \right] \right) + \frac{N^m}{m! n^{m\beta}} \omega_{1,m}^{\max} \left(f_\alpha, \frac{1}{n^\beta} \right) + \left(\frac{2e^4 \|b - a\|_\infty^m \|f_\alpha\|_{\infty,m}^{\max} N^m}{m!} \right) e^{-2n^{(1-\beta)}} \right\},
$$

iv) Suppose $f_\alpha \left(x_0 \right) = 0$, *for all* $\alpha : |\alpha| = 1, ..., m$; $x_0 \in \left(\prod_{i=1}^{N} [a_i, b_i] \right)$. *Then*

$$
|F_n \left(f, x_0 \right) - f \left(x_0 \right)| \leq (4.1488766)^N \cdot
$$
(4.19)

$$
\left\{ \frac{N^m}{m! n^{m\beta}} \omega_1^{\max} \left(f_\alpha, \frac{1}{n^\beta} \right) + \left(\frac{2e^4 \|b - a\|_\infty^m \|f_\alpha\|_{\infty,m}^{\max} N^m}{m!} \right) e^{-2n^{(1-\beta)}} \right\},
$$

notice in the last the extremely high rate of convergence at $n^{-\beta(m+1)}$.

Proof. Consider $g_z(t) := f(x_0 + t(z - x_0))$, $t \geq 0$; $x_0, z \in \prod_{i=1}^{N} [a_i, b_i]$.
Then

$$g_z^{(j)}(t) = \left[\left(\sum_{i=1}^{N} (z_i - x_{0i}) \frac{\partial}{\partial x_i} \right)^j f \right] (x_{01} + t(z_1 - x_{01}), ..., x_{0N} + t(z_N - x_{0N})),$$

for all $j = 0, 1, ..., m$.

We have the multivariate Taylor's formula

$$f(z_1, ..., z_N) = g_z(1) =$$

$$\sum_{j=0}^{m} \frac{g_z^{(j)}(0)}{j!} + \frac{1}{(m-1)!} \int_0^1 (1-\theta)^{m-1} \left(g_z^{(m)}(\theta) - g_z^{(m)}(0) \right) d\theta.$$

Notice $g_z(0) = f(x_0)$. Also for $j = 0, 1, ..., m$, we have

$$g_z^{(j)}(0) = \sum_{\substack{\alpha := (\alpha_1, ..., \alpha_N), \, \alpha_i \in \mathbb{Z}^+, \\ i=1,...,N, \, |\alpha| := \sum_{i=1}^{N} \alpha_i = j}} \left(\frac{j!}{\prod_{i=1}^{N} \alpha_i!} \right) \left(\prod_{i=1}^{N} (z_i - x_{0i})^{\alpha_i} \right) f_\alpha(x_0).$$

Furthermore

$$g_z^{(m)}(\theta) =$$

$$\sum_{\substack{\alpha := (\alpha_1, ..., \alpha_N), \, \alpha_i \in \mathbb{Z}^+, \\ i=1,...,N, \, |\alpha| := \sum_{i=1}^{N} \alpha_i = m}} \left(\frac{m!}{\prod_{i=1}^{N} \alpha_i!} \right) \left(\prod_{i=1}^{N} (z_i - x_{0i})^{\alpha_i} \right) f_\alpha(x_0 + \theta(z - x_0)),$$

$0 \leq \theta \leq 1$.

So we treat $f \in C^m \left(\prod_{i=1}^{N} [a_i, b_i] \right)$.

Thus, we have for $\frac{k}{n}, x \in \left(\prod_{i=1}^{N} [a_i, b_i] \right)$ that

$$f \left(\frac{k_1}{n}, ..., \frac{k_N}{n} \right) - f(x) =$$

$$\sum_{j=1}^{m} \sum_{\substack{\alpha := (\alpha_1, ..., \alpha_N), \, \alpha_i \in \mathbb{Z}^+, \\ i=1,...,N, \, |\alpha| := \sum_{i=1}^{N} \alpha_i = j}} \left(\frac{1}{\prod_{i=1}^{N} \alpha_i!} \right) \left(\prod_{i=1}^{N} \left(\frac{k_i}{n} - x_i \right)^{\alpha_i} \right) f_\alpha(x) + R,$$

where

$$R := m \int_0^1 (1-\theta)^{m-1} \sum_{\substack{\alpha := (\alpha_1, ..., \alpha_N), \, \alpha_i \in \mathbb{Z}^+, \\ i=1,...,N, \, |\alpha| := \sum_{i=1}^{N} \alpha_i = m}} \left(\frac{1}{\prod_{i=1}^{N} \alpha_i!} \right) \left(\prod_{i=1}^{N} \left(\frac{k_i}{n} - x_i \right)^{\alpha_i} \right)$$

$$\cdot \left[f_\alpha \left(x + \theta \left(\frac{k}{n} - x \right) \right) - f_\alpha(x) \right] d\theta.$$

We see that

$$|R| \le m \int_0^1 (1 - \theta)^{m-1} \left(\sum_{|\alpha|=m} \left(\frac{1}{\prod_{i=1}^N \alpha_i!} \right) \left(\prod_{i=1}^N \left| \frac{k_i}{n} - x_i \right|^{\alpha_i} \right) \right. \cdot$$

$$\left| f_\alpha \left(x + \theta \left(\frac{k}{n} - x \right) \right) - f_\alpha(x) \right| \right) d\theta \le m \int_0^1 (1 - \theta)^{m-1} \cdot$$

$$\left(\sum_{|\alpha|=m} \left(\frac{1}{\prod_{i=1}^N \alpha_i!} \right) \left(\prod_{i=1}^N \left| \frac{k_i}{n} - x_i \right|^{\alpha_i} \right) \omega_1 \left(f_\alpha, \theta \left\| \frac{k}{n} - x \right\|_\infty \right) \right) d\theta \le (*).$$

Notice here that

$$\left\| \frac{k}{n} - x \right\|_\infty \le \frac{1}{n^\beta} \Leftrightarrow \left| \frac{k_i}{n} - x_i \right| \le \frac{1}{n^\beta}, \quad i = 1, ..., N.$$

We further observe

$$(*) \le m \cdot \omega_{1,m}^{\max} \left(f_\alpha, \frac{1}{n^\beta} \right) \int_0^1 (1 - \theta)^{m-1} \left(\sum_{|\alpha|=m} \left(\frac{1}{\prod_{i=1}^N \alpha_i!} \right) \left(\prod_{i=1}^N \left(\frac{1}{n^\beta} \right)^{\alpha_i} \right) \right) d\theta =$$

$$\left(\frac{\omega_{1,m}^{\max} \left(f_\alpha, \frac{1}{n^\beta} \right)}{(m!) \, n^{m\beta}} \right) \left(\sum_{|\alpha|=m} \frac{m!}{\prod_{i=1}^N \alpha_i!} \right) = \left(\frac{\omega_{1,m}^{\max} \left(f_\alpha, \frac{1}{n^\beta} \right)}{(m!) \, n^{m\beta}} \right) N^m.$$

Conclusion: When $\left\| \frac{k}{n} - x \right\|_\infty \le \frac{1}{n^\beta}$, we proved

$$|R| \le \left(\frac{N^m}{m! n^{m\beta}} \right) \omega_{1,m}^{\max} \left(f_\alpha, \frac{1}{n^\beta} \right).$$

In general we notice that

$$|R| \le m \int_0^1 (1 - \theta)^{m-1} \left(\sum_{|\alpha|=m} \left(\frac{1}{\prod_{i=1}^N \alpha_i!} \right) \left(\prod_{i=1}^N (b_i - a_i)^{\alpha_i} \right) 2 \|f_\alpha\|_\infty \right) d\theta =$$

$$2 \sum_{|\alpha|=m} \frac{1}{\prod_{i=1}^N \alpha_i!} \left(\prod_{i=1}^N (b_i - a_i)^{\alpha_i} \right) \|f_\alpha\|_\infty \le$$

$$\left(\frac{2 \|b - a\|_\infty^m \|f_\alpha\|_{\infty,m}^{\max}}{m!} \right) \left(\sum_{|\alpha|=m} \frac{m!}{\prod_{i=1}^N \alpha_i!} \right) = \frac{2 \|b - a\|_\infty^m \|f_\alpha\|_{\infty,m}^{\max} N^m}{m!}.$$

We proved in general

$$|R| \leq \frac{2 \|b - a\|_{\infty}^m \|f_\alpha\|_{\infty,m}^{\max} N^m}{m!} := \lambda_3.$$

Next we see that

$$U_n := \sum_{k=\lceil na \rceil}^{\lfloor nb \rfloor} \Theta (nx - k) R =$$

$$\sum_{\substack{k=\lceil na \rceil \\ : \left\| \frac{k}{n} - x \right\|_\infty \leq \frac{1}{n^\beta}}}^{\lfloor nb \rfloor} \Theta (nx - k) R + \sum_{\substack{k=\lceil na \rceil \\ : \left\| \frac{k}{n} - x \right\|_\infty > \frac{1}{n^\beta}}}^{\lfloor nb \rfloor} \Theta (nx - k) R.$$

Consequently

$$|U_n| \leq \left(\sum_{\substack{k=\lceil na \rceil \\ : \left\| \frac{k}{n} - x \right\|_\infty \leq \frac{1}{n^\beta}}}^{\lfloor nb \rfloor} \Theta (nx - k) \right) \frac{N^m}{m! n^{m\beta}} \omega_{1,m}^{\max} \left(f_\alpha, \frac{1}{n^\beta} \right) + e^4 \lambda_3 e^{-2n^{(1-\beta)}}$$

$$\leq \frac{N^m}{m! n^{m\beta}} \omega_{1,m}^{\max} \left(f_\alpha, \frac{1}{n^\beta} \right) + e^4 \lambda_3 e^{-2n^{(1-\beta)}}.$$

We have established that

$$|U_n| \leq \frac{N^m}{m! n^{m\beta}} \omega_{1,m}^{\max} \left(f_\alpha, \frac{1}{n^\beta} \right) + \left(\frac{2e^4 \|b - a\|_{\infty}^m \|f_\alpha\|_{\infty,m}^{\max} N^m}{m!} \right) e^{-2n^{(1-\beta)}}.$$

We observe

$$\sum_{k=\lceil na \rceil}^{\lfloor nb \rfloor} f \left(\frac{k}{n} \right) \Theta (nx - k) - f (x) \sum_{k=\lceil na \rceil}^{\lfloor nb \rfloor} \Theta (nx - k) =$$

$$\sum_{j=1}^{m} \left(\sum_{|\alpha|=j} \left(\frac{f_\alpha (x)}{\prod_{i=1}^{N} \alpha_i!} \right) \left(\sum_{k=\lceil na \rceil}^{\lfloor nb \rfloor} \Theta (nx - k) \left(\prod_{i=1}^{N} \left(\frac{k_i}{n} - x_i \right)^{\alpha_i} \right) \right) \right)$$

$$+ \sum_{k=\lceil na \rceil}^{\lfloor nb \rfloor} \Theta (nx - k) R.$$

The last says that

$$F_n^* (f, x) - f (x) \left(\sum_{k=\lceil na \rceil}^{\lfloor nb \rfloor} \Theta (nx - k) \right) -$$

$$\sum_{j=1}^{m} \left(\sum_{|\alpha|=j} \left(\frac{f_\alpha(x)}{\prod_{i=1}^{N} \alpha_i!} \right) F_n^* \left(\prod_{i=1}^{N} (\cdot - x_i)^{\alpha_i}, x \right) \right) = U_n.$$

Clearly F_n^* is a positive linear operator.

Therefore (here $\alpha_i \in \mathbb{Z}^+ : |\alpha| = \sum_{i=1}^{N} \alpha_i = j$)

$$\left| F_n^* \left(\prod_{i=1}^{N} (\cdot - x_i)^{\alpha_i}, x \right) \right| \le F_n^* \left(\prod_{i=1}^{N} |\cdot - x_i|^{\alpha_i}, x \right) =$$

$$\sum_{k=\lceil na \rceil}^{\lfloor nb \rfloor} \left(\prod_{i=1}^{N} \left| \frac{k_i}{n} - x_i \right|^{\alpha_i} \right) \Theta(nx - k) =$$

$$\sum_{\substack{k=\lceil na \rceil \\ : \left\| \frac{k}{n} - x \right\|_\infty \le \frac{1}{n^\beta}}}^{\lfloor nb \rfloor} \left(\prod_{i=1}^{N} \left| \frac{k_i}{n} - x_i \right|^{\alpha_i} \right) \Theta(nx - k) +$$

$$\sum_{\substack{k=\lceil na \rceil \\ : \left\| \frac{k}{n} - x \right\|_\infty > \frac{1}{n^\beta}}}^{\lfloor nb \rfloor} \left(\prod_{i=1}^{N} \left| \frac{k_i}{n} - x_i \right|^{\alpha_i} \right) \Theta(nx - k) \le$$

$$\frac{1}{n^{\beta j}} + \prod_{i=1}^{N} (b_i - a_i)^{\alpha_i} \left(\sum_{\substack{k=\lceil na \rceil \\ : \left\| \frac{k}{n} - x \right\|_\infty > \frac{1}{n^\beta}}}^{\lfloor nb \rfloor} \Theta(nx - k) \right) \le$$

$$\frac{1}{n^{\beta j}} + \left(\prod_{i=1}^{N} (b_i - a_i)^{\alpha_i} \right) e^4 e^{-2n^{(1-\beta)}}.$$

So we have proved that

$$\left| F_n^* \left(\prod_{i=1}^{N} (\cdot - x_i)^{\alpha_i}, x \right) \right| \le \frac{1}{n^{\beta j}} + \left(\prod_{i=1}^{N} (b_i - a_i)^{\alpha_i} \right) e^4 e^{-2n^{(1-\beta)}},$$

for all $j = 1, ..., m$.

At last we observe

$$\left| F_n(f, x) - f(x) - \sum_{j=1}^{m} \left(\sum_{|\alpha|=j} \left(\frac{f_\alpha(x)}{\prod_{i=1}^{N} \alpha_i!} \right) F_n \left(\prod_{i=1}^{N} (\cdot - x_i)^{\alpha_i}, x \right) \right) \right| \le$$

$$(4.1488766)^N \cdot \left| F_n^*(f, x) - f(x) \sum_{k=\lceil na \rceil}^{\lfloor nb \rfloor} \Theta(nx - k) - \right.$$

$$\sum_{j=1}^{m} \left(\sum_{|\alpha|=j} \left(\frac{f_\alpha(x)}{\prod_{i=1}^{N} \alpha_i!} \right) F_n^* \left(\prod_{i=1}^{N} (\cdot - x_i)^{\alpha_i}, x \right) \right) \Bigg| .$$

Putting all of the above together we prove theorem.   ∎

4.4 Complex Multivariate Neural Network Quantitative Approximations

We make

Remark 4.9. *Let* $X = \prod_{i=1}^{n} [a_i, b_i]$ *or* $\mathbb{R}^N$, *and* $f : X \to \mathbb{C}$ *with real and imaginary parts* $f_1, f_2 : f = f_1 + i f_2$, $i = \sqrt{-1}$. *Clearly* f *is continuous iff* f_1 *and* f_2 *are continuous.*
Given that $f_1, f_2 \in C^m(X)$, $m \in \mathbb{N}$, *it holds*

$$f_\alpha(x) = f_{1,\alpha}(x) + i f_{2,\alpha}(x), \tag{4.20}$$

where α *denotes a partial derivative of any order and arrangement.*
We denote by $C_B(\mathbb{R}^N, \mathbb{C})$ *the space of continuous and bounded functions* $f : \mathbb{R}^N \to \mathbb{C}$. *Clearly* f *is bounded, iff both* f_1, f_2 *are bounded from* $\mathbb{R}^N$ *into* $\mathbb{R}$, *where* $f = f_1 + i f_2$.
Here we define

$$F_n(f, x) := F_n(f_1, x) + i F_n(f_2, x), \quad x \in \left(\prod_{i=1}^{n} [a_i, b_i] \right), \tag{4.21}$$

and

$$\overline{F}_n(f, x) := \overline{F}_n(f_1, x) + i \overline{F}_n(f_2, x), \quad x \in \mathbb{R}^N. \tag{4.22}$$

We see here that

$$|F_n(f, x) - f(x)| \le |F_n(f_1, x) - f_1(x)| + |F_n(f_2, x) - f_2(x)|, \tag{4.23}$$

and

$$\|F_n(f) - f\|_\infty \le \|F_n(f_1) - f_1\|_\infty + \|F_n(f_2) - f_2\|_\infty. \tag{4.24}$$

Similarly we get

$$\left| \overline{F}_n(f, x) - f(x) \right| \le \left| \overline{F}_n(f_1, x) - f_1(x) \right| + \left| \overline{F}_n(f_2, x) - f_2(x) \right|, \quad x \in \mathbb{R}^N, \tag{4.25}$$

and

$$\left\| \overline{F}_n(f) - f \right\|_\infty \le \left\| \overline{F}_n(f_1) - f_1 \right\|_\infty + \left\| \overline{F}_n(f_2) - f_2 \right\|_\infty. \tag{4.26}$$

We give

Theorem 4.10. *Let* $f \in C\left(\prod_{i=1}^{n}[a_i, b_i], \mathbb{C}\right)$, $f = f_1 + if_2$, $0 < \beta < 1$, $n, N \in \mathbb{N}$, $x \in \left(\prod_{i=1}^{n}[a_i, b_i]\right)$. *Then*
i)

$$|F_n(f, x) - f(x)| \leq (4.1488766)^N \cdot \tag{4.27}$$

$$\left\{\omega_1\left(f_1, \frac{1}{n^\beta}\right) + \omega_1\left(f_2, \frac{1}{n^\beta}\right) + 2e^4\left(\|f_1\|_\infty + \|f_2\|_\infty\right)e^{-2n^{(1-\beta)}}\right\} =: \Phi_1,$$

ii)

$$\|F_n(f) - f\|_\infty \leq \Phi_1. \tag{4.28}$$

Proof. Use of Theorem 4.6 and Remark 4.9.   ∎

We present

Theorem 4.11. *Let* $f \in C_B\left(\mathbb{R}^N, \mathbb{C}\right)$, $f = f_1 + if_2$, $0 < \beta < 1$, $n, N \in \mathbb{N}$, $x \in \mathbb{R}^N$. *Then*
i)

$$\left|\overline{F}_n(f, x) - f(x)\right| \leq \omega_1\left(f_1, \frac{1}{n^\beta}\right) + \omega_1\left(f_2, \frac{1}{n^\beta}\right) + \tag{4.29}$$

$$2e^4\left(\|f_1\|_\infty + \|f_2\|_\infty\right)e^{-2n^{(1-\beta)}} =: \Phi_2,$$

ii)

$$\left\|\overline{F}_n(f) - f\right\|_\infty \leq \Phi_2. \tag{4.30}$$

Proof. By Theorem 4.7 and Remark 4.9.   ∎

In the next we discuss high order of complex approximation by using the smoothness of f.

We present

Theorem 4.12. *Let* $f : \prod_{i=1}^{n}[a_i, b_i] \to \mathbb{C}$, *such that* $f = f_1 + if_2$. *Assume* $f_1, f_2 \in C^m\left(\prod_{i=1}^{n}[a_i, b_i]\right)$, $0 < \beta < 1$, $n, m, N \in \mathbb{N}$, $x \in \left(\prod_{i=1}^{n}[a_i, b_i]\right)$. *Then*
i)

$$\left|F_n(f, x) - f(x) - \sum_{j=1}^{m}\left(\sum_{|\alpha|=j}\left(\frac{f_\alpha(x)}{\prod_{i=1}^{N}\alpha_i!}\right)F_n\left(\prod_{i=1}^{N}(\cdot - x_i)^{\alpha_i}, x\right)\right)\right| \leq$$

$$\tag{4.31}$$

$$(4.1488766)^N \cdot \left\{\frac{N^m}{m!n^{m\beta}}\left(\omega_{1,m}^{\max}\left(f_{1,\alpha}, \frac{1}{n^\beta}\right) + \omega_{1,m}^{\max}\left(f_{2,\alpha}, \frac{1}{n^\beta}\right)\right) + \right.$$

$$\left.\left(\frac{2e^4\|b - a\|_\infty^m\left(\|f_{1,\alpha}\|_{\infty,m}^{\max} + \|f_{2,\alpha}\|_{\infty,m}^{\max}\right)N^m}{m!}\right)e^{-2n^{(1-\beta)}}\right\},$$

ii)

$$|F_n(f,x) - f(x)| \leq (4.1488766)^N \cdot \tag{4.32}$$

$$\left\{ \sum_{j=1}^{m} \left(\sum_{|\alpha|=j} \left(\frac{|f_{1,\alpha}(x)| + |f_{2,\alpha}(x)|}{\prod_{i=1}^{N} \alpha_i!} \right) \left[\frac{1}{n^{\beta j}} + \right. \right. \right.$$

$$\left. \left. \left(\prod_{i=1}^{N} (b_i - a_i)^{\alpha_i} \right) \cdot e^4 e^{-2n^{(1-\beta)}} \right] \right) +$$

$$\frac{N^m}{m! n^{m\beta}} \left(\omega_{1,m}^{\max} \left(f_{1,\alpha}, \frac{1}{n^\beta} \right) + \omega_{1,m}^{\max} \left(f_{2,\alpha}, \frac{1}{n^\beta} \right) \right) +$$

$$\left. \left(\frac{2e^4 \|b - a\|_\infty^m \left(\|f_{1,\alpha}\|_{\infty,m}^{\max} + \|f_{2,\alpha}\|_{\infty,m}^{\max} \right) N^m}{m!} \right) e^{-2n^{(1-\beta)}} \right\},$$

iii)

$$\|F_n(f) - f\|_\infty \leq (4.1488766)^N \cdot \tag{4.33}$$

$$\left\{ \sum_{j=1}^{m} \left(\sum_{|\alpha|=j} \left(\frac{\|f_{1,\alpha}(x)\|_\infty + \|f_{2,\alpha}(x)\|_\infty}{\prod_{i=1}^{N} \alpha_i!} \right) \left[\frac{1}{n^{\beta j}} + \right. \right. \right.$$

$$\left. \left. \left(\prod_{i=1}^{N} (b_i - a_i)^{\alpha_i} \right) \cdot e^4 e^{-n^{(1-\beta)}} \right] \right) +$$

$$\frac{N^m}{m! n^{m\beta}} \left(\omega_{1,m}^{\max} \left(f_{1,\alpha}, \frac{1}{n^\beta} \right) + \omega_{1,m}^{\max} \left(f_{2,\alpha}, \frac{1}{n^\beta} \right) \right) +$$

$$\left. + \left(\frac{2e^4 \|b - a\|_\infty^m \left(\|f_{1,\alpha}\|_{\infty,m}^{\max} + \|f_{2,\alpha}\|_{\infty,m}^{\max} \right) N^m}{m!} \right) e^{-2n^{(1-\beta)}} \right\},$$

iv) Suppose $f_\alpha(x_0) = 0$, for all $\alpha : |\alpha| = 1, ..., m$; $x_0 \in \left(\prod_{i=1}^{N} [a_i, b_i] \right)$.
Then

$$|F_n(f, x_0) - f(x_0)| \leq (4.1488766)^N \cdot \tag{4.34}$$

$$\left\{ \frac{N^m}{m! n^{m\beta}} \left(\omega_{1,m}^{\max} \left(f_{1,\alpha}, \frac{1}{n^\beta} \right) + \omega_{1,m}^{\max} \left(f_{2,\alpha}, \frac{1}{n^\beta} \right) \right) + \right.$$

$$\left. \left(\frac{2e^4 \|b - a\|_\infty^m \left(\|f_{1,\alpha}\|_{\infty,m}^{\max} + \|f_{2,\alpha}\|_{\infty,m}^{\max} \right) N^m}{m!} \right) e^{-2n^{(1-\beta)}} \right\},$$

notice in the last the extremely high rate of convergence at $n^{-\beta(m+1)}$.

Proof. By Theorem 4.8 and Remark 4.9.   ∎

References

[1] Anastassiou, G.A.: Rate of convergence of some neural network operators to the unit-univariate case. J. Math. Anal. Appli. 212, 237–262 (1997)

[2] Anastassiou, G.A.: Rate of convergence of some multivariate neural network operators to the unit. J. Comp. Math. Appl. 40, 1–19 (2000)

[3] Anastassiou, G.A.: Quantitative Approximations. Chapman&Hall/CRC, Boca Raton (2001)

[4] Anastassiou, G.A.: Univariate sigmoidal neural network approximation. Journal of Comp. Anal. and Appl. (accepted 2011)

[5] Anastassiou, G.A.: Univariate hyperbolic tangent neural network approximation. Mathematics and Computer Modelling 53, 1111–1132 (2011)

[6] Anastassiou, G.A.: Multivariate hyperbolic tangent neural network approximation. Computers and Mathematics 61, 809–821 (2011)

[7] Chen, Z., Cao, F.: The approximation operators with sigmoidal functions. Computers and Mathematics with Applications 58, 758–765 (2009)

[8] Haykin, S.: Neural Networks: A Comprehensive Foundation, 2nd edn. Prentice Hall, New York (1998)

[9] Mitchell, T.M.: Machine Learning. WCB-McGraw-Hill, New York (1997)

[10] McCulloch, W., Pitts, W.: A logical calculus of the ideas immanent in nervous activity. Bulletin of Mathematical Biophysics 7, 115–133 (1943)